BOOK I

ALGEBRAIC SUDOKU

A Fun Way to Develop, Enhance, and Review Students' Algebraic Skills

Author: Tony G. Williams, Ed.D.
Editors: Howard Brenner and Bonnie Krueger
Design and Layout: Kati Baker

Copyright: 2011 Lorenz Educational Press, a Lorenz company, and its licensors.
All rights reserved.

Permission to photocopy the student activities in this book is hereby granted to one teacher as part of the purchase price. This permission may only be used to provide copies for this teacher's specific classroom setting. This permission may not be transferred, sold, or given to any additional or subsequent user of this product. Thank you for respecting copyright laws.

Printed in the United States of America

ISBN 978-1-4291-2269-6

MILLIKEN
P.O. Box 802 • Dayton, OH 45401
www.LorenzEducationalPress.com

Dear Teacher or Parent,

Welcome to most exciting and beneficial algebraic learning resource on the market today. *Algebraic Sudoku* will teach your child/student algebra in a way never previously accomplished. In this book, the first part in a series of two, you will find 33 Sudoku puzzles, each of which aligns with the standard algebra curriculum. This book is divided into three units:

- Unit I: Foundations
- Unit II: Equations
- Unit III: Linear Equations

Each Sudoku puzzle is like a mini-lesson, with background, discussion, strategy, and demonstration for solving each problem. By solving the algebra problems, students are given enough data that will allow them to reason their way through the remaining cells of the Sudoku puzzle that follows. Each Sudoku puzzle includes a solution key. Every puzzle, along with the problems and mini-lessons associated with it, are presented on a ready-to-use, reproducible master that can be easily copied as transparencies for full-class instruction and discussion.

Algebraic Sudoku is an ideal supplement for students currently enrolled in algebra or students planning to take algebra in the near future. It is an excellent tool for kids and grown-ups to keep their algebraic skills sharp after having completed an algebra course. It is wonderful for home-schoolers as well as students in a traditional classroom. It makes an excellent companion during the summer break. Not only is *Algebraic Sudoku* fun and exciting, it challenges students' minds with exhilarating puzzles that develop logic, reasoning skills, concentration, and confidence.

I hope that you find the *Algebraic Sudoku* puzzles, and the mini-lessons and problems associated with each, the most fun and exciting way to achieve algebra mastery.

Sincerely,

Tony G. Williams, Ed.D.

Dedicated to my parents, Jean and Grady, who are retired math teachers, and to my wife Sharon and our triplets TLC (Tony, Leah, and Christina)

Also, to my brilliant Uncle Charles, and my siblings, Cheryl and Bruce, two committed math educators

TABLE OF CONTENTS

Introduction 4
How It All Works 5
Suggested Strategies 6
Pencil Marks 7

Unit I: Foundations 10
Unit II: Equations 21
Unit III: Linear Equations 30

Answer Key 43

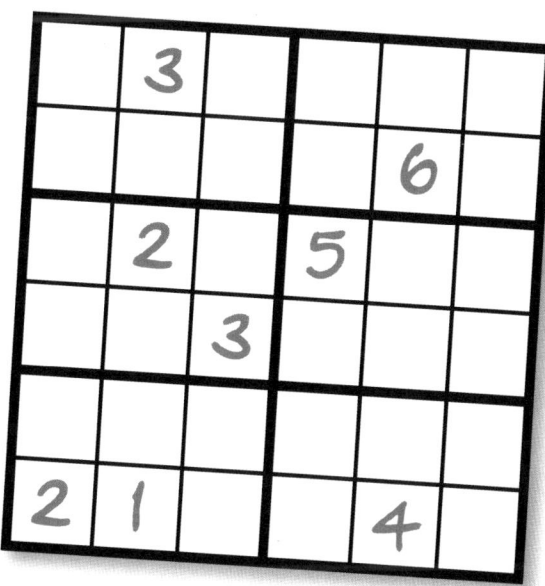

INTRODUCTION

The word *Sudoku* is Japanese—*su* means *number*, and *doku* refers to a single place on a puzzle board. Although its name is Japanese, the game origins are actually European and American. Sudoku puzzles come in a variety of sizes or grids (9x9, 6x6, 5x5, 4x4, etc.). The objective is to fill the grid with digits so that each column, row, and each of the nine 3×3 sub-grids that compose the puzzle contain all of the digits from 1 to 9. (We will discuss strategy later in this section.) Most of the Sudoku puzzles in the book are of the 6x6 variety, with a few 9x9 and 4x4 grids added to the mix.

Today, Sudoku is played all over the world. Many newspapers have added the game alongside the crossword puzzle and, as a result, have reported a jump in circulation. *USA Today*'s most recent list of best-selling books tracked that seven of the top 100 were compilations of Sudoku puzzles. Also, millions of dollars are being made from Sudoku software, games, and online programs. But why? The most obvious reason is that Sudoku is fun; challenging, but fun! There is a sense of accomplishment one receives when successfully completing a puzzle. Sudoku tests one's ability to concentrate, reason, and think logically.

When you combine the excitement of Sudoku with the importance of algebra, you have a winning combination. Students who successfully complete an algebra course are twice as likely to graduate from college as students who lack such preparation (Adelman, 1999; Evan, Gray, and Olchefske, 2006). The majority of employees who earn more than $40,000 a year completed algebra in high school (Achieve, Inc., 2006). A national poll revealed that two-thirds of the students who completed algebra were well-prepared for demands of the workplace (Carnevale and Desrochers, 2003). And yet, there are increasing numbers of students who are not prepared for and fail to successfully complete algebra, as is evident by the vast and growing demand for remedial mathematics education courses among students in four-year colleges and community colleges across the nation. Data shows that 71% of America's degree-granting institutions offer an average of 2.5 remedial courses for skill-deficient students (Business Higher Education Forum, 2005). Overall, these deficiencies are further intensified by factors such as income and race. Research shows that most children from low-income backgrounds enter school with far less knowledge than peers from middle-income backgrounds, and the achievement gap in mathematical knowledge progressively widens throughout their PreK-12 years (National Mathematics Achievement Panel, 2008). However, these achievement gaps can be significantly reduced or even eliminated if low-income and minority students increase their success in high school mathematics and science courses (Evan, Gray, and Olchefske, 2006).

Algebra is essentially both the bridge and the gateway for many students. Algebra can lead students into an exciting world with opportunities awaiting them. We, as math educators, must be as innovative as possible in reaching our students, in enabling our students to reach their fullest potential. I hope this book will be a valuable resource as you strive for success in the classroom.

HOW IT ALL WORKS

This book is divided into three units, with each unit divided into various content areas. For each Sudoku puzzle, there is a specific content area that often includes a mini-lesson that describes the concept and provides an example(s).

The Basics
For each lesson, students must first complete the assigned algebra problems. They will then be asked to either place their answers directly into the Sudoku puzzle, or match their answers and place the corresponding numbers in the Sudoku puzzle. From their efforts, students will be given enough numbers to begin working on the Sudoku puzzle. It is then up to the students to solve the remaining cells of the puzzle.

A system has been devised to easily identify specific cells of a puzzle, using alphabets horizontally across the top and numbers vertically along the left side. Here are some examples:

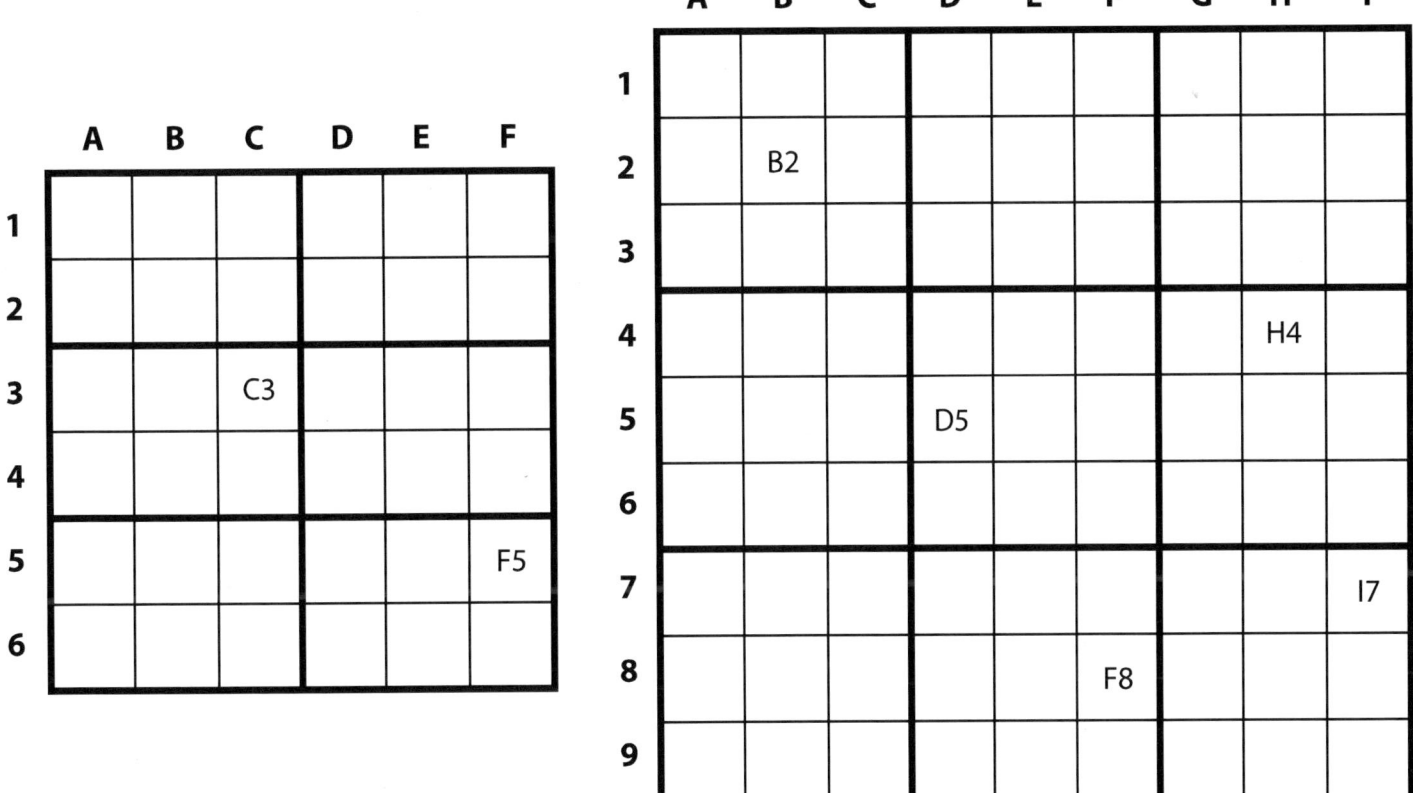

Solutions are provided for each Sudoku puzzle. Whenever possible, classroom discussion should be held to address any remaining questions and to provide further clarification.

Negative Numbers
It is important to note that the cells in these Sudoku puzzles may have negative numbers. Answers with incorrect signs are considered wrong. However, for the purpose of completing the entire grid, a negative number represents a positive number in the counting sequence. For example, a "-2" represents "2". In other words, it is okay to have a sequence such as "-1, 2, -3, 4 -5, 6, -7, -8, 9", provided that these numbers are the correct solutions to the problems presented in the lesson.

SUGGESTED STRATEGIES

Solving Sudoku puzzles should rely more on logic and reasoning, rather than guessing, arithmetic, or any mechanical system. For the novice, it will take some practice and specific strategies; however, the process can be mastered rather easily over a short period of time.

Sudoku Basics
The rules of Sudoku dictate that every cell in the grid is filled with a number. For a 4x4, you would use the numbers 1 to 4; for a 6x6 grid, you would use the numbers 1 to 6; for a 9x9 grid, you would use the numbers 1 to 9; and so on. The restriction is that you can only use each number once in each row, each column, and in each of the box-shaped sub-grids (indicated by bold lines). Also, for the purpose of *Algebraic Sudoku*, a negative number represents a positive number in the counting sequence.

Here are some typical examples of some completed Sudoku puzzles.

	A	B	C	D
1	-1	2	4	3
2	3	-4	-2	1
3	2	1	-3	-4
4	4	3	1	2

	A	B	C	D	E	F
1	4	3	-6	2	-5	1
2	1	-5	2	4	6	-3
3	6	-2	1	5	3	-4
4	5	4	-3	6	1	2
5	-3	-6	4	-1	2	5
6	2	1	5	3	4	-6

	A	B	C	D	E	F	G	H	I
1	1	4	-9	3	-5	2	6	7	8
2	-2	-6	3	1	-7	-8	9	-5	-4
3	-7	8	-5	6	4	9	-3	2	1
4	3	2	-1	9	-8	4	5	6	7
5	6	-7	8	5	2	3	-1	4	-9
6	-5	9	-4	7	-6	1	8	-3	2
7	8	-5	6	-2	9	-7	4	1	-3
8	4	-1	2	8	-3	6	-7	-9	5
9	-9	3	7	4	1	5	2	8	-6

PENCIL MARKS

One of the best strategies to develop students' logic and reasoning skills in solving Sudoku puzzles is by using pencil marks. (Very few players/students can complete Sudoku puzzles by only writing in the final numbers.)

The pencil mark method is an excellent way to get started. Students can learn various shortcuts, patterns, and even their own methods as they advance in their analysis. Let's do an example of the pencil mark method using a 6x6 puzzle. Initially, you will be given the numbers of a few cells to get you started:

	A	B	C	D	E	F
1						
2		5	2			3
3	6				3	
4		4				2
5	3			1	2	
6						

Step 1: First, write in all the possible numbers for the vacant cells. When looking for possible numbers for a vacant cell, you must first look at the given numbers in the same line as that cell both horizontally and vertically, as well as the given numbers in the same sub-grid. For example, for cell A1, vertically we have a 6 and a 3, horizontally there are no numbers in the row, and in its sub-grid we have a 5 and a 2. Therefore, the possible numbers for cell A1 are 1 and 4.

	A	B	C	D	E	F
1	1 4					
2		5	2			3
3	6				3	
4		4				2
5	3			1	2	
6						

Initially, it is recommended that you do this for every vacant cell. As you advance, you may develop shortcuts and other tools for this step. However, as you're learning the process, your first step might look something like this:

	A	B	C	D	E	F
1	1 4	1 3 6	1 3 4 6	2 4 5 6	1 4 5 6	1 4 5 6
2	1 4	5	2	4 6	1 4 6	3
3	6	1 2	1 4 5	4 5	3	1 4 5
4	1 5	4	1 3 5	3 5 6	1 5 6	2
5	3	6	4 5	1	2	4 5 6
6	1 2 4 5	1 2 6	1 4 5 6	3 4 5 6	4 5 6	4 5 6

As they advance, most Sudoku players using pencil marks tend to come up with their own systems to help them complete the grids.

Step 2: The next step is to begin your analysis. You want to look for cells where there are only one or two possible numbers that will fit. In this case, 6 can only work for cell B5, thus eliminating the other 6s horizontally, vertically, and in the same sub-grid. This is done by circling the numbers as shown:

	A	B	C	D	E	F
1	1 4	1 3 ⑥	1 3 4 6	2 4 5 6	1 4 5 6	1 4 5 6
2	1 4	5	2	4 6	1 4 6	3
3	6	1 2	1 4 5	4 5	3	1 4 5
4	1 5	4	1 3 5	3 5 6	1 5 6	2
5	3	⑥ / 6	4 5	1	2	4 5 ⑥
6	1 2 4 5	1 2 ⑥	1 4 5 ⑥	3 4 5 6	4 5 6	4 5 6

You may also choose to cross out the eliminated pencil marks rather than circling them.

8

Step 3: Continue your analysis in much the same way, looking for cells where there are only one or two possible numbers that will fit. Once you properly place a number, be sure to eliminate it as a possibility in that row horizontally and vertically, as well as in the same sub-grid, by circling the number in those locations.

Continuing with the analysis, your next moves might be, but are not necessary restricted to, the following:

- B5: The only number possible for that cell is 6.
- C1: This is the only place for the 6 to go in that sub-grid.
- D1: This is the only place for the 2 to go in that sub-grid.
- B1: This is the only place for the 3 to go in that sub-grid.
- C4: This is the only place for the 3 to go in that sub-grid.
- B3: This is the only place for the 2 to go in that sub-grid.
- B6: This is the only place for the 1 to go in that column.
- A6: This is the only place for the 2 to go in that sub-grid.
- D6: This is the only place for the 3 to go in that sub-grid.
- F6: This is the only place for the 6 to go in that column.
- A4: This is the only place for the 5 to go in that column.
- C3: This is the only number remaining for that sub-grid is 1.
- E4: This is the only place for the 1 to go in that sub-grid.
- D4: This is the only place for the 6 to go in that sub-grid.
- D3: This is the only place for the 5 to go in that column.
- D2: 4 is the last remaining number in that column.
- F3: 4 is the last remaining number in that sub-grid.
- F5: 5 is the last remaining number in that row.
- C5: This is the only place for the 4 to go in that row.
- C6: 5 is the last remaining number in that column.
- E6: 4 is the last remaining number in that row.
- E1: 5 is the last remaining number for that cell.
- E2: 6 is the last remaining number for that cell.
- A2: 1 is the last remaining number in that row.
- A1: 4 is the last remaining number in that column.
- F1: 1 is the last remaining number in the Sudoku puzzle.

Of course, the more Sudoku puzzles that your students solve, the more confident, skillful, and adept they become at finding solutions. However, the primary purpose of this book is for all users to master algebraic skills. The Sudoku puzzles are a fun and exciting way for students to check their answers, evaluate their assignments, and assess their progress. Sudoku adds an additional element to the learning process that involves logic, reasoning, and fun, which is intended to encourage students to work with a greater sense of purpose, enthusiasm, eagerness, excitement, and frequency.

Name _____ Date _____ Puzzle #1

ADDING LIKE AND UNLIKE SIGNS

When adding like signs, simply add the absolute values, and keep the sign for your sum.

Examples: A) 7 + 10 = 17 B) -7 + (-10) = -17

When adding unlike signs, subtract the absolute values of the numbers, and use the sign of the number with greatest absolute value in your answer.

Examples: A) -7 + 16 = 9 B) 7 + (-16) = -9

Directions: Add, and place the sums in the appropriate cells of the Sudoku grid. Then solve the puzzle.

(A1) -3 + (-4) =	(C8) $\frac{1}{2}$ + (-6 $\frac{1}{2}$) =	(H2) - 4 $\frac{3}{5}$ + (-2 $\frac{2}{5}$) =
(A4) 2.7 + 2.3 =	(D3) 17 + (-14) =	(H3) 100 + (-98) =
(A9) -1 + 0 =	(D6) -3.929 + -0.071 =	(H6) -5.72 + 10.72 =
(B1) -8 $\frac{7}{8}$ + $\frac{7}{8}$ =	(D9) -19 + 10 =	(H9) 0 + (-6) =
(B4) -4.28 + 5.28 =	(F1) -4 $\frac{1}{5}$ + (-$\frac{4}{5}$) =	(I1) -9 $\frac{3}{4}$ + 5 $\frac{3}{4}$ =
(B7) -2 + 7 =	(F4) 30 + (-23) =	(I3) 51 + (-50) =
(B8) 4.2 + (-0.2) =	(G2) 6.2 + (-1.2) =	(I5) -13.3 + 4.3 =
(B9) -20 + 13 =	(G3) 17 + (-8) =	(I6) -9.7 +11.7 =
(C4) 4.35 + 4.65 =	(G6) 0.55 + 0.45 =	(I8) -1.68 + (-1.32) =
(C7) 21 + (-23) =	(H1) 34 + (-37) =	(I9) 33 + (-38) =

Name _____ Date _____ Puzzle #2

SUBTRACTING SIGNED NUMBERS

Subtracting is the same as adding the opposite (or additive inverse) of the number being subtracted.

Examples: A) $-5 - 10 = -5 + (-10) = -15$ B) $12 - (-23) = 12 + 23 = 35$

Directions: Subtract, and place the differences in the appropriate cells of the Sudoku grid. Then solve the puzzle.

(A5) $7 - 10 =$	(D5) $15 - 20 =$	(G1) $-2.5 - (-7.5) =$
(A7) $-3 - 6 =$	(E1) $-6 - 3 =$	(G6) $18 - 24 =$
(B1) $4 - (-4) =$	(E2) $0 - (-3) =$	(G7) $-5 - 2 =$
(B4) $-9 - (-4) =$	(E3) $0.77 - 2.77 =$	(H2) $-29 - (-20) =$
(B5) $5 - (-1) =$	(E4) $0 - 1 =$	(H4) $-3 - 5 =$
(B6) $-\frac{7}{8} - \frac{1}{8} =$	(E6) $-3\frac{7}{12} - 3\frac{5}{12} =$	(H5) $-4\frac{3}{5} - (-6\frac{3}{5}) =$
(B8) $-5.2 - 1.8 =$	(E7) $19 - 14 =$	(H6) $0 - (-4) =$
(C3) $23 - 30 =$	(E8) $-18 - (-24) =$	(H9) $6.1 - 11.1 =$
(C4) $-24 - (-20) =$	(E9) $-36 - (-32) =$	(I3) $-26 - (-32) =$
(C9) $-1 - 1 =$	(F5) $-1 - 3 =$	(I5) $-48 - (-55) =$
(D2) $-4 - (-8) =$	(F8) $-7.25 - 1.75 =$	(I8) $49 - 53 =$

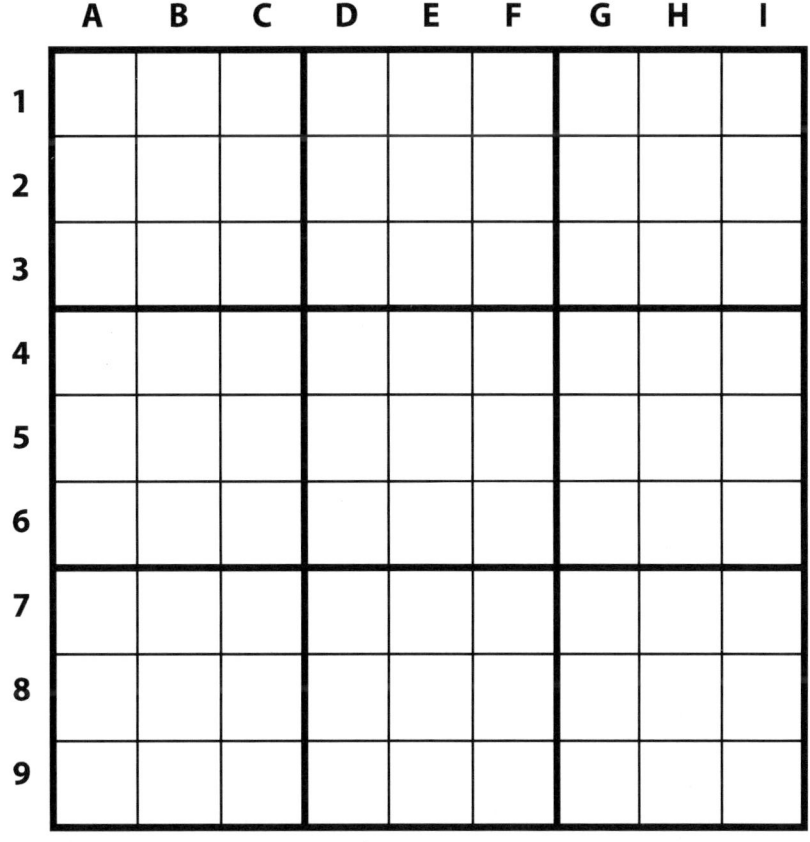

Foundations

11

Name _____ Date _____ Puzzle #3

MULTIPLYING AND DIVIDING SIGNED NUMBERS

When multiplying or dividing *rational numbers* (negative and positive whole numbers, fractions, decimals, etc.), the rules are easy to remember. When multiplying or dividing two unlike signs (- + or + -) the product or quotient is always **negative**. When multiplying or dividing like signs (+ + or - -) the product or quotient is always **positive**.

Here are a few examples:

A) $-6 \cdot 3 = -18$ B) $-15 \div (-5) = 3$ C) $24 \div (-6) = -4$ D) $-7 \cdot -9 = 63$

Directions: Multiply or divide, and place the products or quotients in the appropriate cells of the Sudoku grid. Then solve the puzzle.

(A4) $-35 \div (-7) =$	(D5) $-27 \div 3 =$	(F6) $111 \div (-37) =$
(A8) $42 \div (-6) =$	(D7) $-5 \cdot -1 =$	(G1) $-63 \div 21 =$
(A9) $-3 \cdot 3 =$	(D8) $33 \div (-11) =$	(G4) $-7 \cdot (-1) =$
(B1) $-14 \cdot -\frac{1}{2} =$	(E1) $-\frac{2}{3} \cdot -6 =$	(G6) $\frac{3}{5} \div -\frac{3}{25} =$
(B5) $8 \cdot -\frac{1}{8} =$	(E3) $\frac{3}{5} \div (-\frac{1}{5}) =$	(G7) $-4 \div (-\frac{1}{2}) =$
(B9) $72 \div (-9) =$	(E5) $-2 \cdot (-3) =$	(H1) $-0.1 \cdot 10 =$
(C3) $-81 \div (-9) =$	(E7) $99 \div (-11) =$	(H5) $28 \div (-7) =$
(C4) $24 \cdot -0.25 =$	(E9) $-1.75 \cdot 4 =$	(H8) $-0.4 \cdot 5 =$
(C6) $-48 \div 6 =$	(F2) $56 \div (-7) =$	(I1) $100 \div (-20) =$
(C9) $-13 \div (-13) =$	(F3) $-175 \div (-25) =$	(I2) $-42 \div 6 =$
(D4) $-\frac{1}{4} \cdot 32 =$	(F5) $-25 \cdot -\frac{1}{5} =$	(I6) $-4 \div (-\frac{2}{3}) =$

12

Foundations

Name _____ Date _____ Puzzle #4

REVIEW OF RATIONAL NUMBER OPERATIONS AND ORDER OF OPERATIONS

Remember the order of operations:
(1) Perform the operations inside the parentheses first;
(2) Simplify the exponents;
(3) Multiply or divide depending on which occurs first from left to right, and
(4) Add or subtract depending on which occurs first from left to right.

Directions: Perform the indicated operation and/or simplify the expression, and place the answers in the appropriate cells of the Sudoku grid. Then solve the puzzle.

(A1) $11 - 3 \cdot 4 =$	(D2) $-2 + 3 \cdot 2 + 4 =$	(F6) $41 + (-37) =$
(A5) $-4 + 11 =$	(D4) $-42 \div -7 =$	(G2) $0 - (-7) =$
(A8) $-6 - (-12) =$	(D5) $5^0 =$	(G4) $4 - (5 + 2) =$
(B1) $(3-5)^2 - 11 =$	(D7) $4 \div (2+2) - 10 =$	(G7) $35 \div (-7) =$
(B4) $-7 \cdot -\frac{1}{7} =$	(D9) $3 \div -\frac{3}{5} =$	(G8) $-\frac{4}{7} \cdot -\frac{7}{4} =$
(B5) $8 - 12 =$	(E2) $22 - 23 =$	(G9) $2^2 =$
(C1) $-12 + 6 =$	(E5) $(-3)^2 =$	(H5) $-3 \cdot 3 + 1 =$
(C2) $6 \div 2 \cdot 3 - 7 =$	(E8) $(2-5)^2 \div 3 + 1 =$	(H6) $-41 + 42 =$
(C3) $-12 \cdot -\frac{1}{3} =$	(F1) $18 \cdot -0.5 =$	(H9) $54 \div (-9) =$
(C6) $-45 \div -9 =$	(F4) $-2 \div 0.25 =$	(I2) $-8 \cdot (-0.75) =$
(C8) $(-2)^3 =$	(F5) $-34 + 36 =$	(I5) $-27 - (-32) =$

Foundations

13

Name _____ Date _____ Puzzle #5

ABSOLUTE VALUE

Absolute value is the unit value a number is from zero on the number line. The symbol for absolute value is two vertical lines (| |).

Examples: |8| = 8 |-8| = 8

Note that the absolute value of any number is always greater than or equal to zero.

Directions: Evaluate the following expressions, and place the answers in the appropriate cells of the Sudoku grid. Then solve the puzzle.

(A4) | -7| =	**(D7)** | 1 | =	**(G8)** | -8 | =
(A8) | 4 | =	**(D8)** | -2 – 3 | =	**(H2)** | -14 ÷ 2 | =
(A9) | 2 – 3| =	**(E2)** | -3 | + | -2 | =	**(H4)** | 4 • -2 | =
(B1) | -4 | =	**(E4)** | -23 + 14 | =	**(H8)** -| 3 | =
(B2) | 6 | =	**(E6)** -5 + | -4 | =	**(H9)** | 5 | + | 4 | =
(B6) | -4 + 2 | =	**(E8)** | -24 +18 | =	**(I1)** | 0 – 3 | =
(B8) | 3 – (-6) | =	**(F2)** | -8 | – 12 =	**(I2)** | -17 | – 26 =
(C2) | -3 | – | 4 | =	**(F3)** | 2 | =	**(I5)** | -33 + 31 | =
(C5) | 0 – 9 | =	**(F6)** - | -3 | =	**(I6)** | -28 • -0.25 | =
(C9) | 1.3 – 8.3 | =	**(G1)** -8.8 + | -3.8 | =	**(I9)** | 7 ÷ -1 $\frac{3}{4}$ | =
(D4) | -$\frac{3}{5}$ | + | 1 $\frac{2}{5}$ | =	**(G5)** 6 $\frac{4}{7}$ – | -2 $\frac{4}{7}$ | =	

14   Foundations

Name _____ Date _____ Puzzle #6

COLLECTING LIKE TERMS

Like terms have the same variable(s) and exponent(s). The terms *2xy* and *-4xy* are like terms, but x^2 and x^3 are not. Like terms may be collected by adding or subtracting the coefficients in front of the terms.

For example: $8x - 3x + x = 6x$

Directions: Simplify the expressions below, and find your answer in the box on the right. Place the numbers in the appropriate cells of the Sudoku grid, and then solve the puzzle.

(A2) $2x + y - x + 3y =$
(A3) $-4x - 2y + x - y =$
(B3) $xy - 3xy =$
(B6) $-5xy + xy - y =$
(C4) $5x + 4y - 3x^2 - xy =$
(C6) $-5x - y - 3x + 2y =$
(D1) $5x - 5xy - x - xy =$
(D3) $2x^2 - x^2 - y^2 + 3x^2 =$
(E1) $3x^3 - 2x^2 + x + 2x^2 =$
(E4) $-4x^2 - 4xy - y + xy - x^2 =$
(F4) $6x^2 - y^2 - 5x^2 + y^2 =$
(F5) $x^2y^2 - x^2y^3 + 3x^2y^2 + x^2y^3 =$

(1) $-3x - 3y$
(3) $5x + 4y - 3x^2 - xy$
(5) $-8x + y$
(4) $3x^3 + x$
(5) $4x^2 - y^2$
(1) $-4xy - y$
(2) $4x^2y^2$
(1) x^2
(2) $-5x^2 - 3xy - y$
(5) $x + 4y$
(2) $4x - 6xy$
(6) $-2xy$

	A	B	C	D	E	F
1						
2						
3						
4						
5						
6						

Foundations

Name _____ Date _____ Puzzle #7

MULTIPLYING TERMS WITH EXPONENTS

When multiplying exponents, multiply the coefficients and add the exponents of the like bases (same variable).

For example: $2x^2 \cdot -4x^3 = -8x^5$

Directions: Multiply the problems below, and find your answer in the box on the right. Place the numbers in the appropriate cells of the Sudoku grid, and then solve the puzzle.

(A2)	$x^2 \cdot 3x^3 =$
(B2)	$-4x \cdot (-2x^2) =$
(B3)	$x^2 \cdot x =$
(B4)	$xy \cdot 3xy =$
(B5)	$3x^2 \cdot 2y^2 =$
(C3)	$-x^4 \cdot 4x =$
(C4)	$4x^3 \cdot (-0.5x^2) =$
(D3)	$-3y \cdot 2y^2 \cdot (-y^3) =$
(D4)	$2x^2y \cdot (-3xy^2) =$
(E2)	$-\tfrac{3}{4}x^4 \cdot (-12x) =$
(E3)	$-4x^2y^2z \cdot 2xy^3z^3 =$
(E4)	$x^2 \cdot y^2 \cdot z^2 =$
(E5)	$3x^5y^6 \cdot (-xy) \cdot 2x^2y =$
(F5)	$-3y^2 \cdot xy^2 \cdot (-2x^2y) =$

(4)	$8x^3$
(5)	$6x^2y^2$
(3)	$6x^3y^5$
(2)	$-6x^3y^3$
(5)	$-8x^3y^5z^4$
(6)	$9x^5$
(3)	$x^2y^2z^2$
(1)	$3x^5$
(3)	$-4x^5$
(6)	$3x^2y^2$
(4)	$6y^6$
(4)	$-6x^8y^8$
(1)	x^3
(4)	$-2x^5$

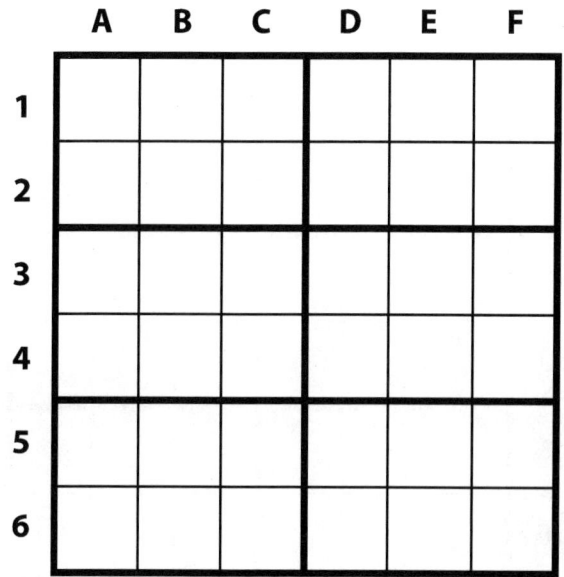

16　　　　　　　　　　　　　　　　　　　　　　　　　　　Foundations

Name _____ Date _____ Puzzle #8

POWER TO A POWER

When an exponent is raised to a power, first raise the coefficient to that power. Then multiply the exponent by that power.

For example: $(2x^3)^3 = 8x^9$

Directions: Simplify the problems below, and find your answer in the box on the right. Place the numbers in the appropriate cells of the Sudoku grid, and then solve the puzzle.

(A1) $(x^2)^4 =$
(A3) $(4x^3)^3 =$
(B1) $(3x^4)^3 =$
(B4) $(-5x^3y)^2 =$
(C4) $(4xy^2)^2 =$
(C5) $(-x^2yz^2)^3 =$
(D2) $(-2x^3y^5)^5 =$
(D3) $(-3x^2y^2z^3)^3 =$
(E3) $(2.5x^3)^2 =$
(E6) $(3x)^2 \cdot (2x)^2 =$
(F4) $(-\frac{1}{2}x^2)^2 =$
(F6) $(5x^4y^2)^0 =$

(3) $64x^9$
(6) $-x^6y^3z^6$
(4) $-27x^6y^6z^9$
(1) $36x^4$
(4) $16x^2y^4$
(6) $\frac{1}{4}x^4$
(5) $25x^6y^2$
(2) 1
(4) $27x^{12}$
(2) $6.25x^6$
(6) x^8
(2) $-32x^{15}y^{25}$

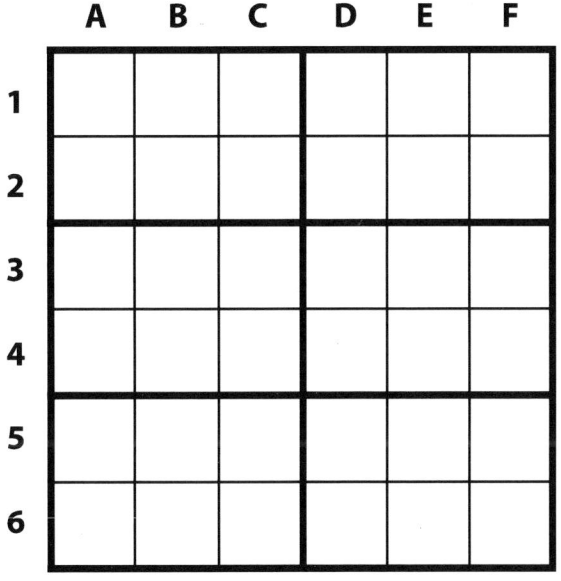

Foundations

17

Name _____ Date _____ Puzzle #9

DIVIDING TERMS WITH EXPONENTS

To divide terms in exponential form with like bases, simplify and subtract the exponents.

By definition, $\dfrac{x^5}{x^2} = x^3$ and $\dfrac{x^2}{x^5} = \dfrac{1}{x^3}$.

For example, $\dfrac{x^6 y^3}{x y^4} = \dfrac{x^5}{y}$

Directions: Divide the terms below, and find your answer in the box on the right. Place the numbers in the appropriate cells of the Sudoku grid, and then solve the puzzle.

(A3) $\dfrac{x^6}{x^4} =$	(D4) $\dfrac{24x^3 y^5}{-8x^3 y^2} =$	(3) x^6	(1) $x^4 y$
(A6) $\dfrac{x^7}{x} =$	(D5) $\dfrac{-15x^4}{-12x^7} =$	(4) -1	(6) x^2
(B4) $\dfrac{x^2}{x^6} =$	(E1) $\dfrac{-x^3 y^4 z^5}{x^3 y^4 z^5} =$	(1) $\dfrac{x}{y^3}$	(4) $\dfrac{5}{4x^3}$
(B6) $\dfrac{x^5 y^4}{x y^3} =$	(E3) $\dfrac{-10xy^5}{2x^4 y^3} =$	(3) $5x^2$	(1) $-\dfrac{5y^2}{x^3}$
(C2) $\dfrac{x^3 y^2}{x^2 y^5} =$	(F1) $\dfrac{-9x^6 y^3}{-9x^5 y^4} =$	(4) $-\dfrac{3}{xy^2}$	(5) $\dfrac{x}{y}$
(C3) $\dfrac{10x^5}{2x^3} =$	(F4) $\dfrac{-63x^5 y^7 z^4}{21x^6 y^9 z^4} =$	(2) $\dfrac{1}{x^4}$	(3) $-3y^3$

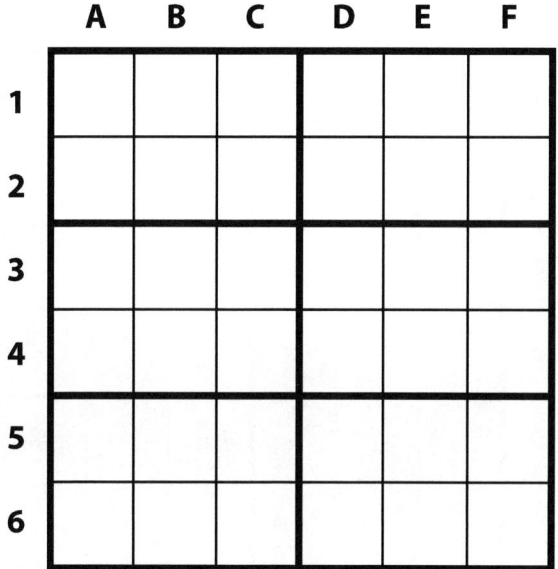

18   Foundations

THE DISTRIBUTIVE PROPERTY

According to the distributive property, $5(x + y) = 5x + 5y$. In other words, multiply each term inside the parentheses by the term outside the parentheses.

Directions: Simplify the expressions below, and find your answer in the box on the right. Place the numbers in the appropriate cells of the Sudoku grid, and then solve the puzzle.

(A2) $3(x + y) =$	**(2)** $-6x + 2y$
(A4) $2(x - y) =$	**(1)** $8x + 12y$
(B4) $4(2x + 3y) =$	**(5)** $-12x^2 + 3xy^2$
(B6) $-2(3x - y) =$	**(6)** $-3x^3 + x^2 - xy$
(C2) $-\frac{1}{2}(2x - 4y) =$	**(3)** $3x + 3y$
(C3) $-3x(4x - y^2) =$	**(2)** $2x - 2y$
(D4) $3y^2(-2x - y + 2xy) =$	**(1)** $10x - 8y$
(D5) $-(2x^2 + y - z) =$	**(5)** $-8x^2 + 6xy$
(E1) $(4x - 3y)(-2x) =$	**(1)** $15xy^3 - 3x^3y^5$
(E3) $3xy^2(5y - x^2y^3) =$	**(5)** $-2x^2 - y + z$
(F3) $x(-3x^2 + x - y) =$	**(2)** $-x + 2y$
(F5) $2(2x - 3y) - 2(y - 3x) =$	**(4)** $-6xy^2 - 3y^3 + 6xy^3$

	A	B	C	D	E	F
1						
2						
3						
4						
5						
6						

Foundations

Name _____ Date _____ Puzzle #11

OVERVIEW OF FOUNDATIONS

Directions: Find the value of each expression by substituting *2* for *x*. Place your answers in the appropriate cells of the Sudoku grid, and then solve the puzzle.

(A1) $x^3 - 1 =$	(D3) $-3x + 2x - x =$	(G1) $-(-x) =$
(A3) $1 - x =$	(D5) $-x + 5 =$	(G3) $5x^2 - x - 4x^2 + 2x + 1 =$
(A7) $7 - (-x) =$	(D6) $(x + 1)^2 =$	(G6) $-2x + 7 =$
(B1) $\dfrac{-18}{x} =$	(E2) $\dfrac{x^5}{x^4} =$	(G8) $\dfrac{6x^7}{2x^6} =$
(B2) $-x - 1 =$	(E3) $12 \div (4 + x) - 11 =$	(H1) $x^0 =$
(B8) $2 + x \cdot 3 =$	(E7) $x^2 - 1 =$	(H2) $-x - 7 =$
(C1) $2(x - 4) =$	(E9) $5x^2 - 4x^2 =$	(H6) $-2 + 2x =$
(C4) $-1 + x =$	(F6) $5(3x - 7) =$	(I1) $12 \cdot \dfrac{x^3}{x^4} =$
(C5) $5 - (-2x) =$	(F8) $x + (-11) =$	(I4) $\dfrac{x^9}{x^6} =$
(D2) $4x^2 - 3x^2 + 3 =$	(F9) $-x - 1 + 5x =$	(I7) $(x^2 + 4)^0 =$

20   Foundations

Name _____ Date _____ Puzzle #12

EQUATION STRATEGY

In order to solve basic one-step equations, first isolate the variable by canceling out the other terms on the same side of the equation. This is done by performing the opposite operation. Division and multiplication are opposite operations. Addition and subtraction are opposite operations.

Examples:
A) In the equation $x + 5 = 12$, subtract 5 from both of the equation to solve, thus the answer is $x = 7$.
B) In the equation $7x = 21$, divide both sides of the equation by 7, giving an answer of $x = 3$.

Directions: Mentally solve the one-step equations. Find your answer in the box on the right. Place the numbers in the appropriate cells of the Sudoku grid, and then solve the puzzle.

(A2) $x - 5 = 6$	(E9) $y + 10 = 31$	(8) -8	(1) 24
(A5) $x + 8 = 12$	(F2) $-5x = 25$	(2) -11	(6) 4
(A8) $y - 8 = 2$	(F3) $8 + x = 20$	(3) 3	(4) -1
(B1) $y + 3 = 9$	(F5) $19 = x + 2$	(2) 11	(6) -9
(B3) $x + 5 = -3$	(F7) $-2x = -44$	(8) -4	(1) 23
(B4) $x - 2 = -1$	(F8) $x - 17 = 3$	(3) 17	(4) 10
(C2) $5x = 15$	(G3) $-6 = x - 6$	(5) 33	(5) -20
(C3) $6y = 42$	(G4) $-x = 20$	(6) -2	(5) 7
(C4) $8x = 16$	(G6) $-3x = 30$	(8) -5	(6) 20
(C6) $-7x = 21$	(G7) $x + 2 = 1$	(9) 12	(5) 18
(C7) $4x = -36$	(G8) $x + 29 = 1$	(8) -15	(8) -10
(D2) $\frac{x}{3} = 8$	(H6) $\frac{x}{5} = -6$	(4) 13	(1) 2
(D3) $x - 11 = -13$	(H7) $34 = 11 + z$	(2) 1	(5) 21
(D5) $-y + 30 = 9$	(H9) $24 = 9 - x$	(4) -3	(9) -7
(D7) $-3x = 33$	(I2) $-8 = x - 21$	(7) 22	(1) 21
(D8) $10y = -40$	(I5) $2x + x = -21$	(3) -30	(7) -28
(E1) $\frac{x}{2} = 9$	(I8) $\frac{x}{-3} = -11$	(3) 0	(4) 6

Equations

21

Name _____ Date _____ Puzzle #13

TWO-STEP EQUATIONS

When solving two-step equations, start by canceling out the number being added to or subtracted from the variable. Then cancel the number being multiplied by or divided into the variable.

For example:

Cancel out the 2 by subtracting it from both sides.
$$3x + 2 = 14$$
$$3x + 2 - 2 = 14 - 2$$
$$3x = 12$$

Then divide both sides by 3. Cancel out the 3, leaving x by itself.
$$\frac{3x}{3} = \frac{12}{3}$$
$$x = 4$$

Directions: Showing all steps, solve the following problems. Place your answers in the appropriate cells of the Sudoku grid, and then solve the puzzle.

(A2) $5x + 8 = 23$ **(A3)** $3x - 9 = 6$ **(B4)** $11x + 7 = 29$ **(B6)** $10y - 12 = 18$

(C3) $5x + 25 = 5$ **(C6)** $2x - 7 = -17$ **(D1)** $49 = 8x + 1$ **(D4)** $-2x + 4 = 6$

(E1) $7 + 6x = 13$ **(E3)** $\frac{x}{3} + 4 = 5$ **(F4)** $\frac{4x}{5} - 3 = 1$ **(F5)** $-23 - 4x = -27$

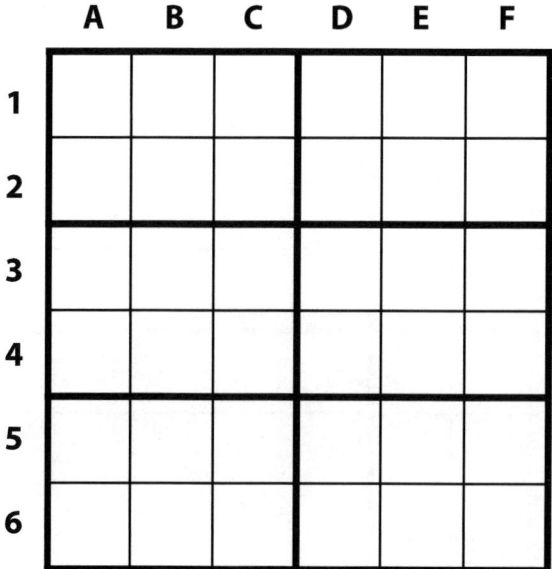

22

Equations

Name _____ Date _____ Puzzle #14

EQUATIONS AND THE DISTRIBUTIVE PROPERTY

From your knowledge of the distributive property, you know that $7(x-3) = 7x - 21$.

Directions: For the problems below, use the distributive property to remove the parentheses. Next, showing all steps, solve each problem. Place your answers in the appropriate cells of the Sudoku grid, and then solve the puzzle.

(A2) $2(x + 3) = 14$

(A5) $4(x - 1) = 4$

(B4) $5(y + 3) = 45$

(B6) $-3(x + 5) = -27$

(C1) $4(x - 2) = -20$

(C3) $6(x + 10) = 72$

(D4) $3(2x + 1) = 33$

(D6) $4(3x - 12) = -24$

(E1) $2(1 + 3x) = 32$

(E3) $-(x - 2) = 8$

(F2) $-45 = 9(5x - 10)$

(F5) $-2(-4x - 9) = -30$

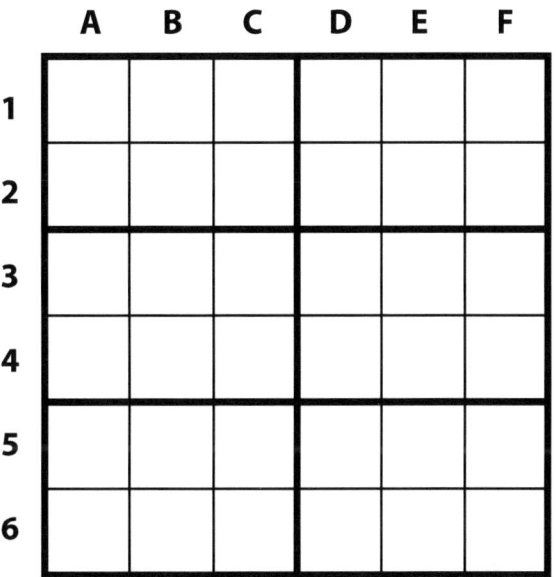

Equations

23

MULTI-STEP EQUATIONS

When solving multi-step equations, first simplify the side of the equation with the variable. Then solve the equation.

For example:
$4(3x - 2) - 2x = 42$
$12x - 8 - 2x = 42$
$10x - 8 = 42$
$10x - 8 + 8 = 42 + 8$
$\frac{10x}{10} = \frac{50}{10}$
$x = 5$

Directions: Showing all steps, solve the following problems. Place your answers in the appropriate cells of the Sudoku grid, and then solve the puzzle.

(A2) $5(x + 1) - 6 = 14$ **(B3)** $3(x - 2) + 5 = 11$ **(C4)** $4(x + 3) - 7 = 29$

(D1) $2(2x - 7) + 3x = 14$ **(D3)** $-3(5x - 10) - x = -66$ **(E4)** $-5x - 2 + 3x - 5 = -17$

(F3) $-3x - 6 + 8 - 2x = -8$ **(F5)** $5(2x - 4) - 4(3x + 8) = -40$

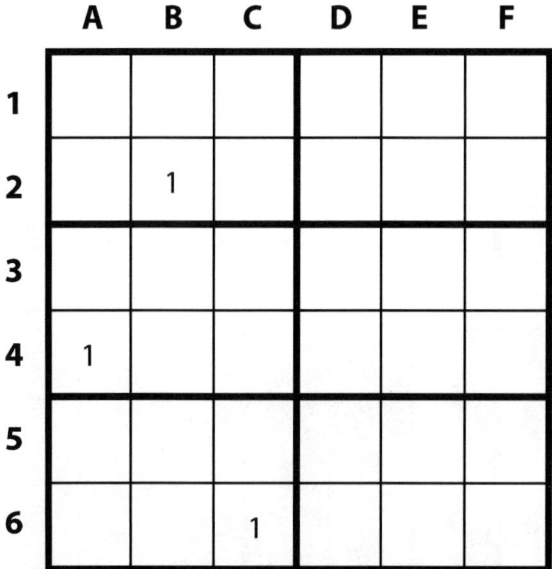

24 Equations

Name _____ Date _____ Puzzle #16

EQUATIONS WITH THE VARIABLE ON BOTH SIDES

When solving equations with variables on both sides, first cancel out the variable on one side.

In the equation $5x - 4 = 14 - 4x$, it is possible to cancel the $5x$ or $-4x$. It might be easier to cancel the $-4x$ by adding $4x$ to both sides, leaving $9x$ on one side. Once one of the variables is canceled, the equation can be solved.

For example:
$$5x - 4 = 14 - 4x$$
$$5x - 4x - 4 = 14 - 4x + 4x$$
$$9x - 4 = 14$$
$$9x - 4 + 4 = 14 + 4$$
$$9x = 18$$
$$x = 2$$

Directions: Showing all steps, solve the following problems. Place your answers in the appropriate cells of the Sudoku grid, and then solve the puzzle.

(A1) $6x - 4 = 41 - 3x$ **(A3)** $5x + 1 = 17 + x$ **(B5)** $-4x + 15 = 6x - 35$

(C3) $5(x - 3) = -2x + 20$ **(D3)** $-2x - 18 = 4x$ **(D6)** $-4(x - 1) = 2(x - 4)$

(E2) $7x - 4 + 3x = 8x + 2$ **(F4)** $-x = -2x + 5$ **(F6)** $-2(-4x + 1) = -5 + 7x$

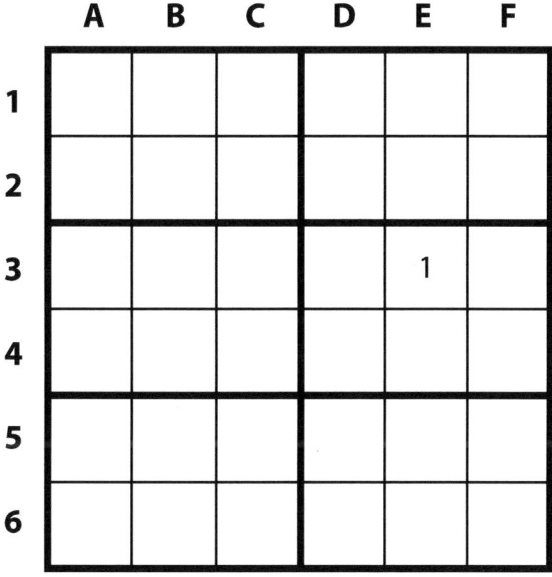

Equations

25

Name _____   Date _____   Puzzle #17

EQUATIONS WITH FRACTIONAL COEFFICIENTS

When solving equations with fractional coefficients, multiply the equation by the least common denominator (LCD) of the fractions.

For example:

$$\frac{2}{3}x + \frac{1}{6} = \frac{1}{4}$$

The LCD is 12, which is the smallest number that 3, 6, and 4 can all divide into evenly. By multiplying by 12, the equation can be simplified to $8x + 2 = 3$, which is more manageable.

$$(12)\frac{2}{3}x + (12)\frac{1}{6} = (12)\frac{1}{4} \longrightarrow 8x + 2 = 3$$

Directions: Multiply by the LCD. Then, showing all steps, solve the following problems. Place your answers in the appropriate cells of the Sudoku grid, and then solve the puzzle.

(A3) $\frac{1}{2}x + \frac{1}{3} = \frac{4}{3}$

(B3) $\frac{2}{3}x - \frac{1}{6} = \frac{5}{9}x + \frac{1}{6}$

(C2) $\frac{5}{6}(2x - 14) + \frac{1}{4}x = -\frac{3}{8}x - \frac{5}{2}$

(D2) $-\frac{2}{5}x - \frac{1}{10}x = \frac{4}{15}(x - 13) + 5$

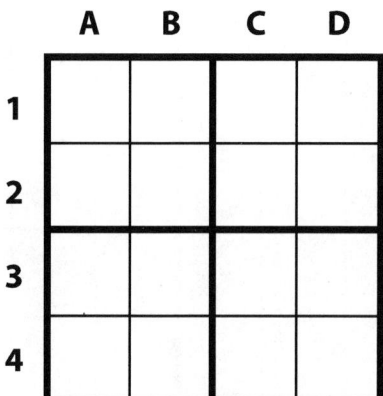

26

Equations

Name _____ Date _____ Puzzle #18

MORE MULTI-STEP EQUATIONS

Directions: Showing as many steps as necessary, solve the following problems. Place your answers in the appropriate cells of the Sudoku grid, and then solve the puzzle.

(A3) $-x = -5$

(A6) $|5 - 9| = x$

(B6) $-4x - 2 = -14$

(C1) $3(7 - x) = 2x - 4$

(C4) $\dfrac{1}{2}x - 4 = \dfrac{3}{4}x - 5$

(D3) $-8x - 6 + x - 8 = -50 - x$

(D6) $0.7(2x + 10) = 0.2(11x + 15)$

(E1) $-|4| = x$

(F1) $7x - 9x + 2 = 0$

(F4) $\dfrac{1}{2}x - \dfrac{3}{7}(4x - 1) = 3x - 8$

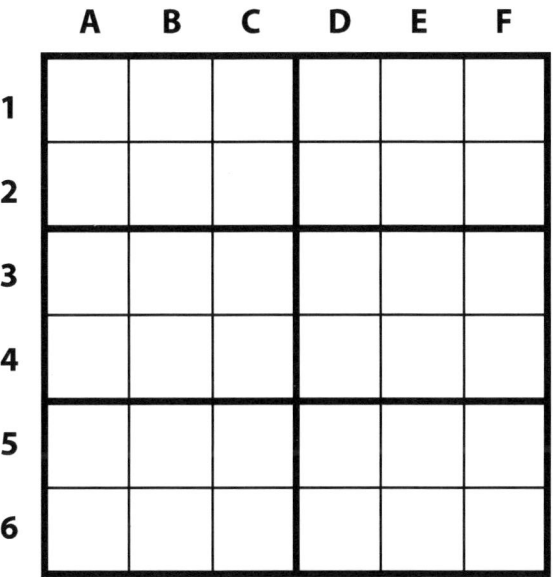

Equations

27

Name _____ Date _____ Puzzle #19

SOLVING INEQUALITIES

When solving inequalities, the steps are exactly the same as solving equations. However, when multiplying or dividing by a negative number, the direction of the inequality must be reversed.

For example, to solve the inequality -5x ≥ 25, divide the sides of the inequality by -5. In doing so, the direction of the inequality is reversed, which gives an answer of x ≤ -5.

Directions: Showing all steps, solve the following inequalities. Find your answer in the box on the right. Place the numbers in the appropriate cells of the Sudoku grid, and then solve the puzzle.

(A3) -7x ≤ 14

(A6) 2x − 8 > -12

(B6) 3x − 5 < 13 − 6x

(2) x < -1
(2) x ≥ 2
(6) x ≥ -1
(6) x ≤ -2
(4) x > 2
(6) x < 2
(2) x ≤ 2
(1) x < -2
(5) x > -2
(3) x ≥ -2

(C1) -4(x +3) ≥ 2x − 24

(C4) $\frac{-x}{2} > 1$

(D3) -4x − 7 ≥ -1 − x

(D6) $x < -\frac{1}{2}$ (2)

(E1) -7x + 2 < -8 − 2x

(F1) 0.7x + 4 ≥ -0.3x + 3

(F4) $\frac{1}{6}x - 2 \geq -\frac{5}{3}$

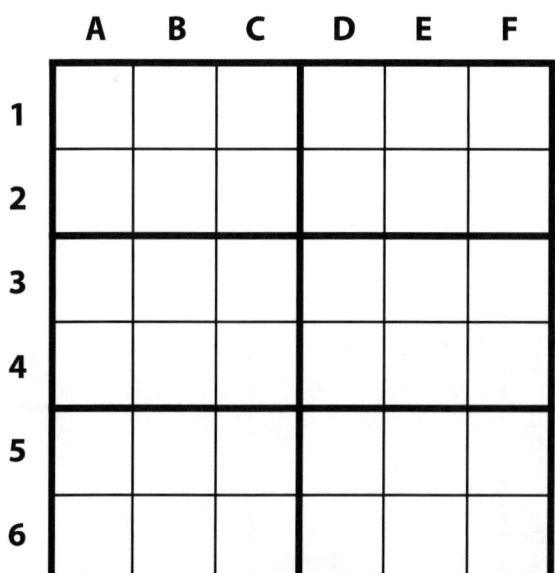

28

Equations

Name _____ Date _____ Puzzle #20

CONSECUTIVE INTEGERS AND OTHER APPLICATIONS

Here are some basic hints for translating word problems into algebraic equations.

- Use the distributive property when dealing with a number times a sum or difference. For example, five times the sum of a number and three can be written as $5(x+3)$.

- When solving problems to determine consecutive integers (e.g. 1, 2, 3), use x to represent the first, $x+1$ for the second, $x+2$ for the third, and so on. For even or odd integers, use x, then $x+2$, then $x+4$, and so on.

Directions: Write an equation for the following word problems. Then solve the equations. Find your answer in the box on the right. Place the numbers in the appropriate cells of the Sudoku grid, and then solve the puzzle.

(A1) Three times a number plus four times the same number minus eight equals forty-one. What is the number?

(B1) Find three consecutive integers whose sum is 105.

(C3) If six times the sum of a number and five equals sixteen times the number, find the number.

(D1) Find three consecutive even integers whose sum is -6.

(E6) Mary wants to average 126 for 3 games of bowling. If she bowls scores of 105 and 132 on her first two games, what score must she bowl on her third game to obtain a 126 average?

(F3) Find three consecutive odd integers whose sum is -27.

(6) -13, -9, -5
(1) 141
(5) 8
(2) 7
(4) 3
(3) 4
(5) -11, -9, -7
(1) -8, -2, 4
(3) 34, 35, 36
(5) -4, -2, 0
(2) 139

	A	B	C	D	E	F
1						
2						
3						
4	6			3		
5						
6			5			3

Equations

29

Name _____ Date _____ Puzzle #21

RECTANGULAR COORDINATE SYSTEMS

A rectangular coordinate system consists of an x-axis (horizontal) and y-axis (vertical) and has four quadrants.

Directions: (I.) Name the coordinates of the indicated points. (II.) Answer the questions related the graph, and place the numbers in the appropriate cells of the Sudoku grid. Then solve the puzzle.

I. G **(B1, F6)**; H **(B4, D1)**; I **(C4, E1)**; J **(A4, E2)**; K **(B2, C5)**; L **(A1, B6)**; M **(E3, F4)**

II. **(E6, D5)** What are the coordinates of the point at which line I and II intersect?
    (F5) In what quadrant is the x-coordinate positive and the y-coordinate negative?
    (D4) In what quadrant are both the x-coordinate and y-coordinate negative?

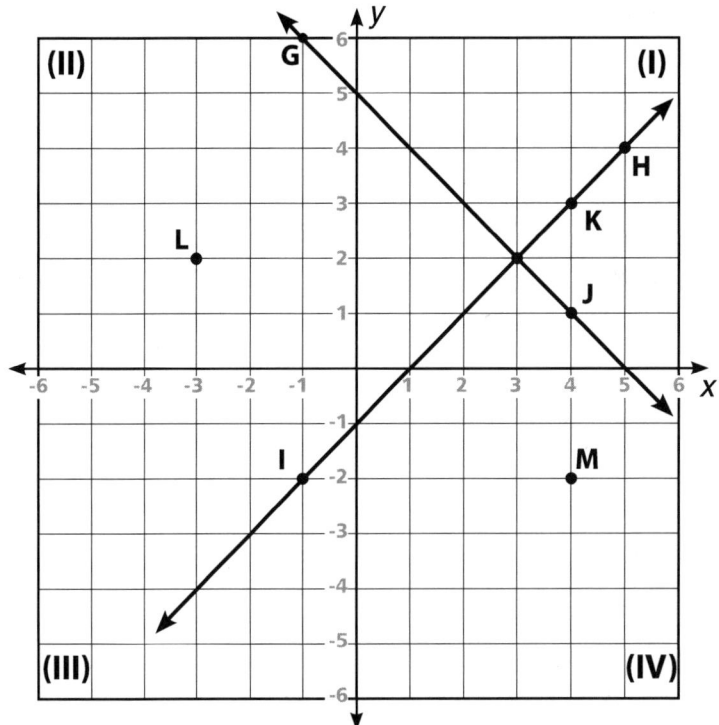

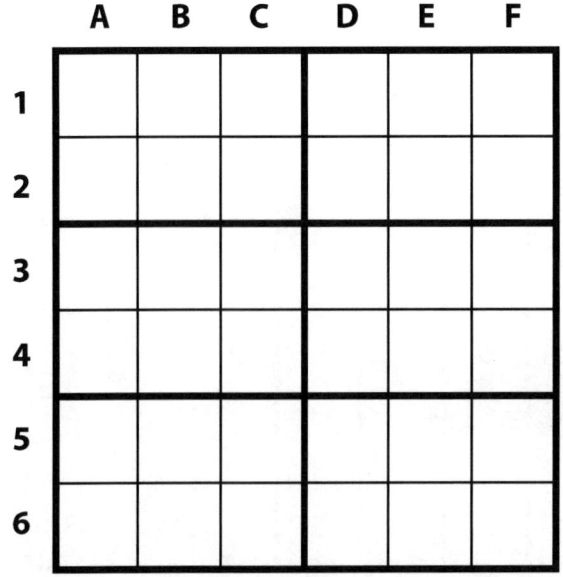

30   *Linear Equations*

Name _____ Date _____ Puzzle #22

GRAPHING LINEAR EQUATIONS

One method of graphing the solution of a linear equation is to substitute a value for *x* or *y*, and then solve for the other variable.

Directions: For each problem, complete the table. Use the points to graph the line for each equation. Place your answers in the appropriate cells of the Sudoku grid, and then solve the puzzle.

I. $x + y = 6$

x	y	
2		(B1)
(C2)	3	
-1		(I5)
(G8)	-2	
0		(C3)

II. $2x = y$

x	y	
1		(B2)
(C9)	4	
4		(A1)
(H3)	-8	
$\frac{1}{2}$		(G4)

III. $x - y = 1$

x	y	
6		(A5)
(B5)	7	
8		(C6)
(H8)	-2	
-3		(F6)

IV. $-2x = y - 3$

x	y	
0		(F1)
(D6)	7	
5		(D4)
(H9)	-3	
$-\frac{1}{2}$		(I7)

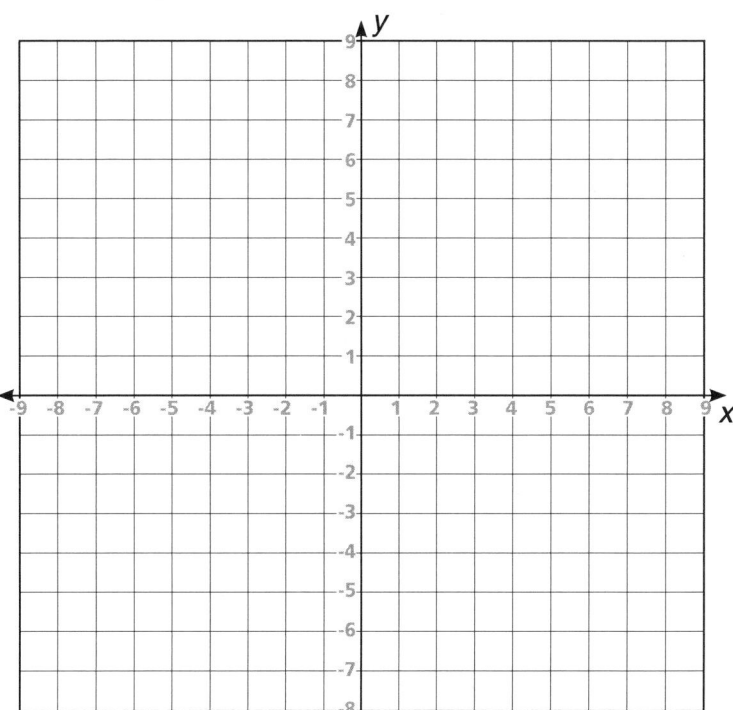

	A	B	C	D	E	F	G	H	I
1							2		
2									
3	7				2				
4	2					9			
5					1			2	
6									5
7		7			3		5		
8						7			
9	4	1		5					6

Linear Equations

31

Name _____ Date _____ Puzzle #23

SLOPE = RISE/RUN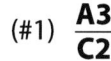

The slope is the steepness of a line. The slope is really a ratio of how many units a line rises (or declines) for each unit it runs (moves to left or right).

To find the slope of a line on a graph, first locate any two points on the line. Count from one point to the next, first by going up (positive) or down (negative), and then right (positive) or left (negative).

For example, let's find the slope of the line in the graph to the right. If you move from point A to point B, you must go up (+) 2 units then to the right (+) 2 units, giving a slope of $\frac{2}{2}$, or 1. If you move from point B to point A, you must first go down (–) 2 units then move to the left (–) 2 units, giving a slope of a $\frac{-2}{-2}$ which is also equal to $\frac{2}{2}$. (It is important to note that vertical lines have no slope, and horizontal lines have slopes equal to zero.)

Directions: Find the slopes of the six lines in the graph. Place the ratios in the appropriate cells of the Sudoku grid, and then solve the puzzle.

(#1) $\frac{A3}{C2}$

(#2) $\frac{A5}{E5}$

(#3) $\frac{E3}{D5}$

(#4) $\frac{B2}{E6}$

(#5) $\frac{F4}{0}$ = no slope

(#6) $\frac{0}{B4}$ = 0 slope

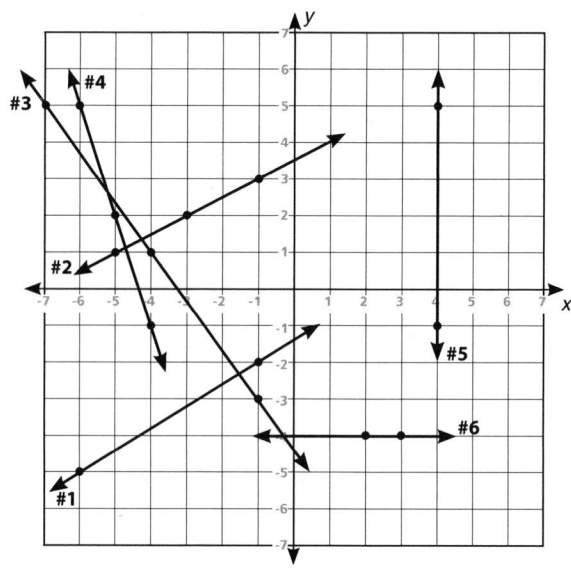

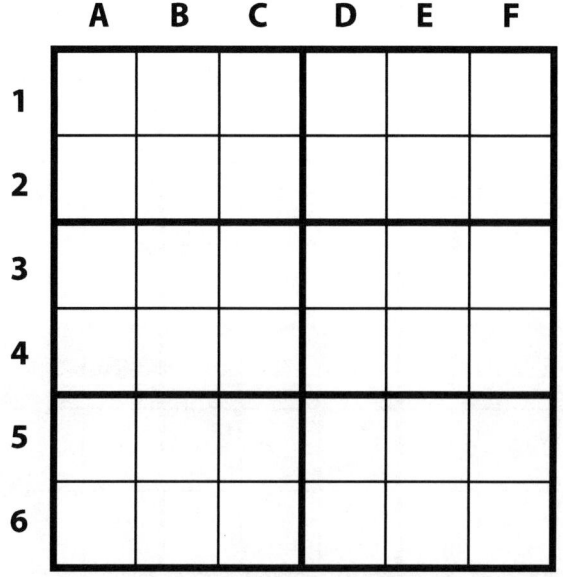

32

Linear Equations

Name _____ Date _____ Puzzle #24

Y-INTERCEPT

The y-intercept of a line is where it crosses the y-axis. In algebra, *b* represents the y-intercept, and *m* represents the slope.

For example:
In Line #1 in the graph below, the slope (m) = $-\frac{1}{3}$ and the y-intercept (b) = -2.

Directions: Determine the slopes and y-intercepts for lines 2 through 5 on the graph below. Place the corresponding numbers in the appropriate cells of the Sudoku grid, and then solve the puzzle.

Line #2: $m = \frac{D3}{B1}$, $b = $ ___ **(C4)**

Line #3: $m = \frac{A3}{F4}$, $b = $ ___ **(C1)**

Line #4: $m = \frac{A6}{D5}$, $b = $ ___ **(F6)**

Line #5: $m = \frac{A2}{E6}$, $b = $ ___ **(D6)**

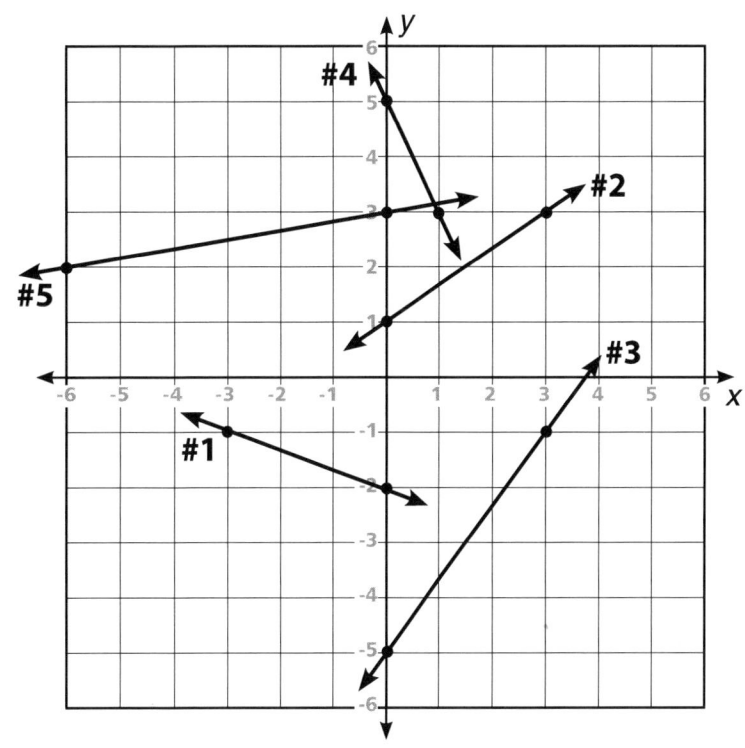

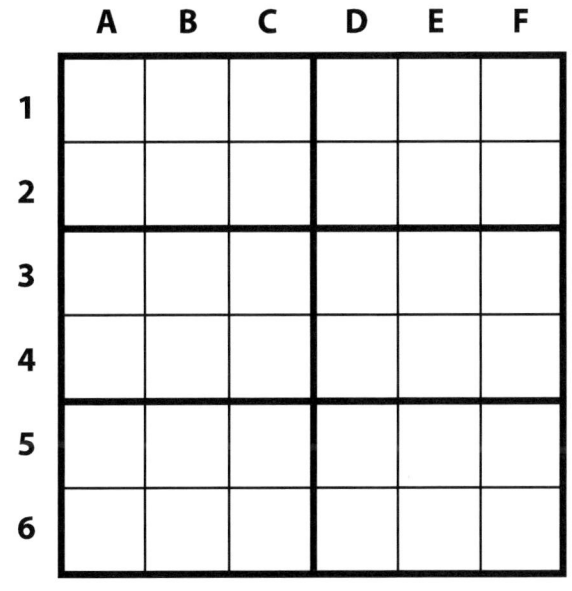

Linear Equations

33

Name _____   Date _____   Puzzle #25

SLOPE FORMULA

One way of finding the slope of a line is by using the following formula:

$$m = \frac{y_2 - y_1}{x_2 - x_1}$$

Choose two points on the line. Designate one point as (x_1, y_1) and the other as (x_2, y_2). Then substitute the values into the formula, and solve.

Let's use the line with the points (2, -3) and (-5, 1) as an example.

$$m = \frac{1 - (-3)}{-5 - 2} = \frac{4}{-7} = -\frac{4}{7}$$

Directions: Find the slope of each line using the two points given. Place the corresponding numbers in the appropriate cells of the Sudoku grid, and then solve the puzzle.

(6,2) (8,7); $\frac{E1}{D6}$ 	(5,3) (10,2); $\frac{A1}{C4}$ 	(-2,-2) (2,1); $\frac{C2}{F6}$

(6,1) (1,7); $\frac{E4}{B6}$ 	(2,6) (-1,2); $\frac{B3}{D3}$ 	(0,-2) (4,0); $\frac{F2}{C1}$

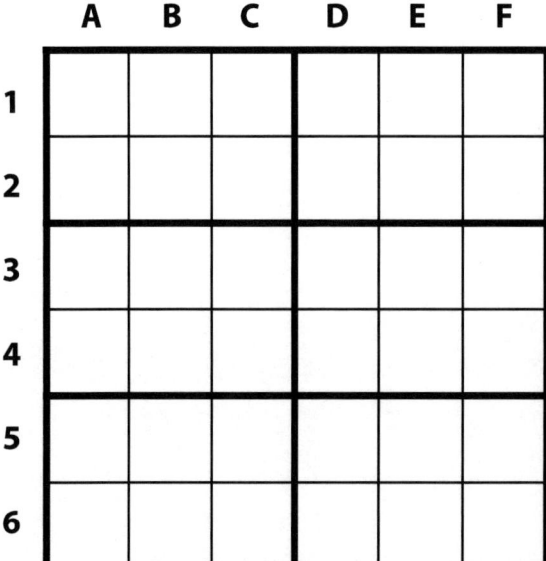

34   *Linear Equations*

Name _____ Date _____ Puzzle #26

GRAPHING THE LINE

Directions: Given the slope and the y-intercept (or a point on the line), find the matching line. Place the corresponding line numbers in the appropriate cells of the Sudoku grid, and then solve the puzzle.

(E4) $m = \dfrac{2}{3}, b = -4$

(A2) $m = -2, b = 5$

(D3) $m = -\dfrac{1}{2}, b = 0$

(C4) $m = -\dfrac{1}{3}, b = -4$

(F5) $m = 2, b = 5$

(D5) $m = \dfrac{1}{5}, b = 0$

(B5) (-5,4), no slope

(C2) $m = 0, b = 2$

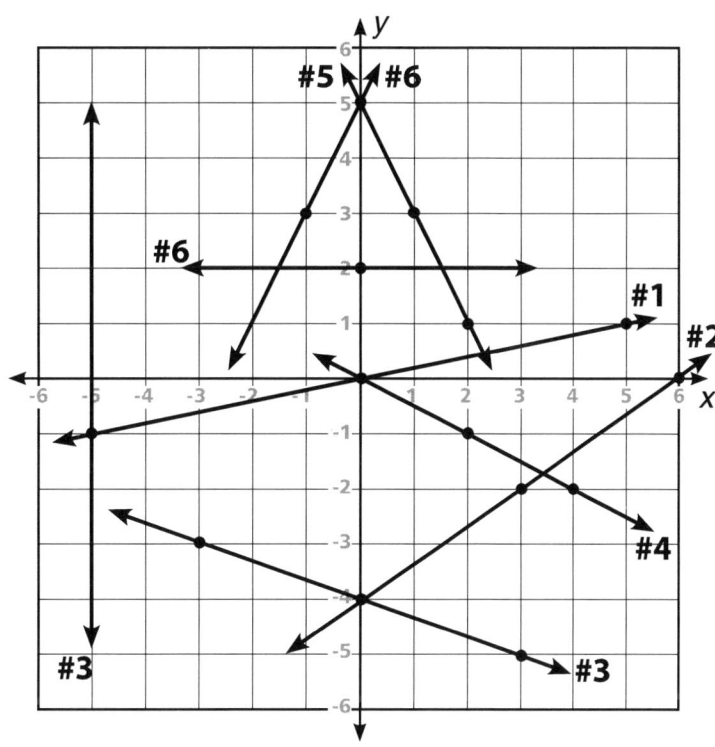

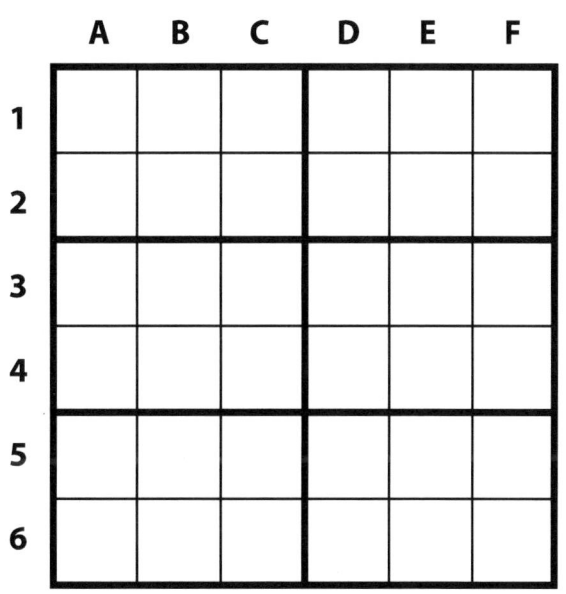

Linear Equations

35

Name _____ Date _____ Puzzle #27

WRITING AN EQUATION

The slope-intercept equation for a line is **y = mx + b**, with *m* being the slope and *b* being the *y*-intercept. For line A in the graph below, the *y*-intercept is 2 and the slope is $\frac{1}{3}$. Thus, the equation for line A is $y = \frac{1}{3}x + 2$.

Directions: Match the equations with lines 1 through 6 on the graph. Place the corresponding line numbers in the appropriate cells of the Sudoku grid, and then solve the puzzle.

(A4) $y = \frac{1}{2}x - 3$ (B6) $y = x + 5$ (C3) $y = -3x$

(D1) $y = -\frac{1}{3}x - 3$ (D4) $y = -\frac{4}{3}x + 5$ (F3) $y = -7$

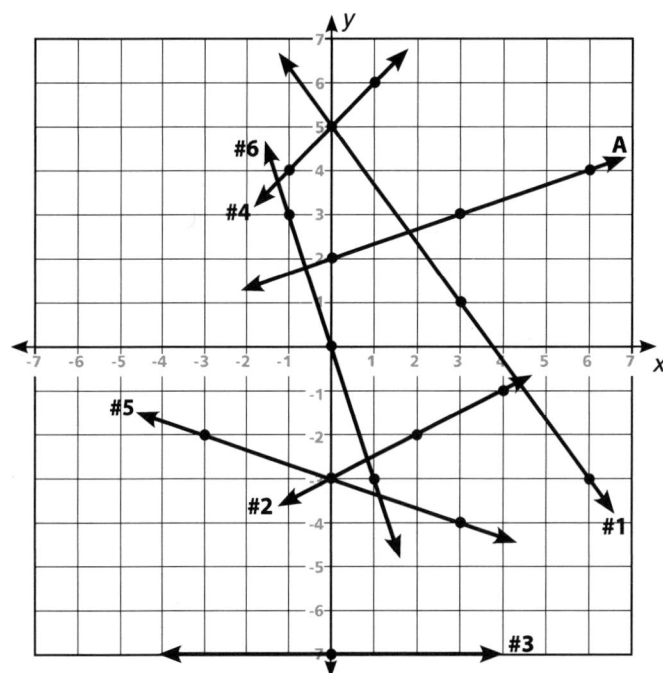

	A	B	C	D	E	F
1	6				2	
2						
3						
4						
5						
6			2			6

Linear Equations

Name _____   Date _____   Puzzle #28

SLOPE-INTERCEPT FORM

When writing an equation in slope-intercept form ($y = mx + b$), m represents the slope and b represents the y-intercept. If solving for y in the equation $2x + 5y = -25$, the resulting equation will be in slope-intercept form.

For example:
$$2x + 5y = -25$$
$$2x - 2x + 5y = -2x - 25$$
$$\frac{5y}{5} = \frac{-2x}{5} - \frac{25}{5}$$

$y = -\frac{2}{5}x - 5$, thus $m = -\frac{2}{5}$ and $b = -5$.

Directions: Write the following equations in slope-intercept form. Identify the slope and y-intercept, and place the corresponding numbers in the appropriate cells of the Sudoku grid. Then complete the puzzle. Show all work!

$2y = 2x + 10$, $m = \dfrac{D1}{B6}$, $b = $ **F4**

$2y + 3x = 12$, $m = \dfrac{C6}{E1}$, $b = $ **A4**

$3x - 6y = -30$, $m = \dfrac{F3}{D3}$, $b = $ **C3**

$\dfrac{1}{2}y + 2 = \dfrac{1}{12}x$, $m = \dfrac{A2}{D6}$, $b = $ **F6**

$y = 4$, slope $= 0$, find y-intercept **(A1)**

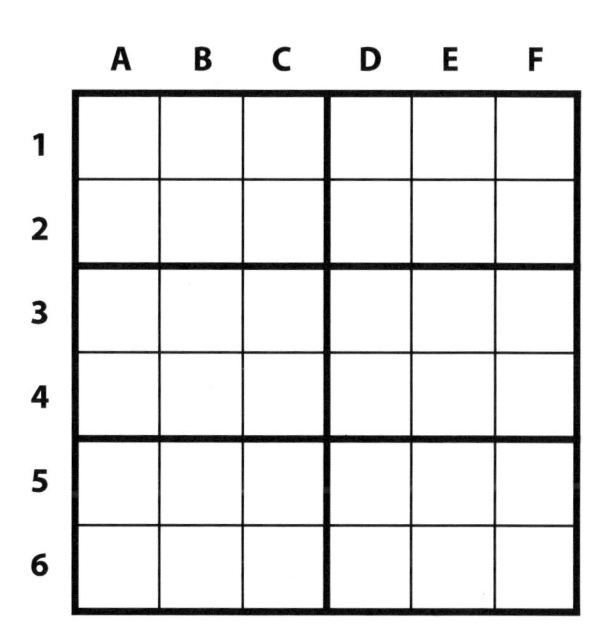

	A	B	C	D	E	F
1						
2						
3						
4						
5						
6						

Linear Equations

Name _____   Date _____   Puzzle #29

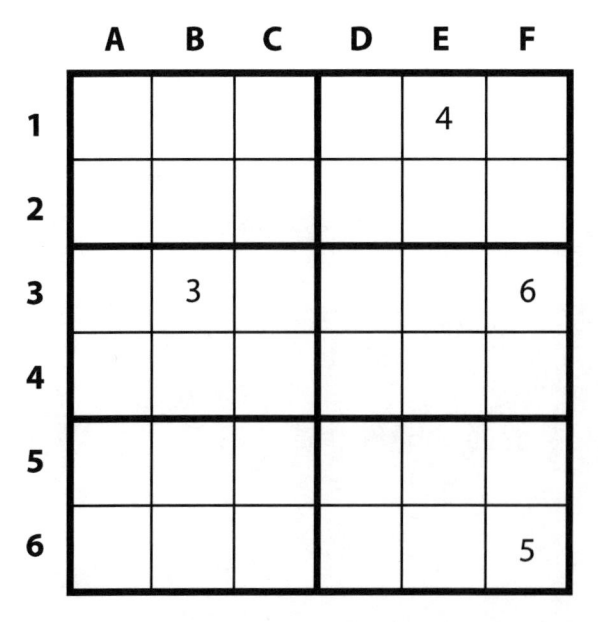

PARALLEL, PERPENDICULAR, AND COINCIDENTAL LINES

Two **parallel lines** have the same slopes. Two lines that are **perpendicular** to one another (forming a right angle) have negative reciprocal slopes, such as $\frac{-4}{3}$ and $\frac{3}{4}$. **Coincidental lines** are essentially the same line, occupying the same space. Coincidental lines have the same slope and the same y-intercept. All other pairs of lines with different slopes are called **intersecting lines**.

Directions: Identify the relationship of the pair of equations given. Use the following rules to fill in the appropriate cells of the Sudoku grid. Then complete the puzzle.

parallel lines ⟶ 4
perpendicular lines ⟶ 3
coincidental lines ⟶ 2
all other intersecting lines ⟶ 1

(A1) $y = -\frac{2}{3}x + 4$
$y = \frac{3}{2}x - 7$

(A4) $y = 2x - 7$
$6x - 3y = 9$

(B2) $y = -\frac{3}{5}x + 1$
$5y - 5 = -3x$

(C1) $4 = 3x - y$
$-3x - y = -4$

(D6) $y = -2x - 3$
$2y = x + 8$

(E4) $4y = 12x - 8$
$9x - 3y = 6$

	A	B	C	D	E	F
1					4	
2						
3		3				6
4						
5						
6						5

38   *Linear Equations*

Name _____ Date _____ Puzzle #30

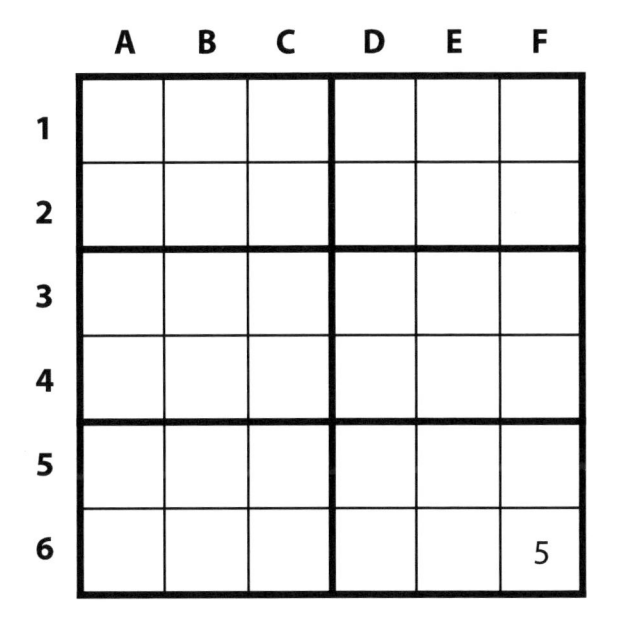

STANDARD FORM: Ax + By = C

The standard form of a line is another way of writing the equation of a line: $Ax + By = C$

Standard form has the following criteria: (1) A, B, and C are integers (positive or negative whole numbers); (2) there are no fractions or decimals in standard form; and (3) the Ax term is positive.

For example, here is how to convert $\frac{1}{2}y = -2x + 1$ to standard form:

First multiply by 2 (LCM) to cancel out the fraction: $y = -4x + 2$
Then add 4x to both sides, leaving: $4x + y = 2$
Notice that the Ax term (4x) is positive.

Directions: Write the following linear equations in standard form. Place the numbers for A, B, and C in the appropriate cells of the Sudoku grid, and then solve the puzzle.

$3x + 6y = 4$ $\frac{1}{2}x = y + 1$ $-2y = -\frac{2}{3}x - \frac{4}{3}$

(A1)x + (A4)y = (B1)   (C2)x + (C4)y = (D5)   (D4)x + (E6)y = (F3)

	A	B	C	D	E	F
1						
2						
3						
4						
5						
6						5

Linear Equations

39

Name _____ Date _____ Puzzle #31

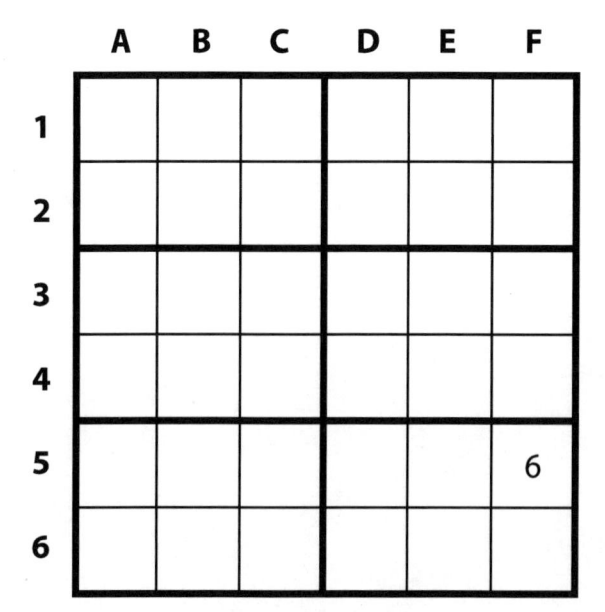

POINT–SLOPE FORM

Point–slope form, $y - y_1 = m(x - x_1)$, can be used to write an equation when only the slope of the line and one point on the line are given. The subscripts represent the known point on the line.

Example:
Write an equation with a slope of 2 that contains the point (-3, 4) on the line.
  $y - 4 = 2(x - (-3))$
  $y - 4 = 2(x + 3)$
  $y - 4 = 2x + 6$
   $y = 2x + 10$

Directions: Given the slope and a point on the line, use point–slope form to find the equation of the line. Find your answer in the box on the right. Place the numbers in the appropriate cells of the Sudoku grid, and then solve the puzzle. Show as much work as possible!

(A3) $m = 3, (5, 2)$ **(A5)** $m = 3, (2,6)$ **(B2)** $m = 3, (-1,-4)$

(4) $y = 3x - 13$
(3) $y = 3x - 5$
(5) $y = 3x$
(4) $y = 3x - 1$
(5) $y = 3x + 24$
(2) $y = 3x - 6$
(1) $y = 2$
(4) $y = 3x - 16$
(5) $y = 3x - 17$
(6) $x = 2$

(B4) $m = 3, (5,-2)$ **(C5)** $m = 3, (0, -5)$ **(D2)** $m = 3, (-8, 0)$

(E1) $m = 0, (1,2)$ **(E5)** $m = 3, (1, -13)$ **(F2)** $m = 3, (3,3)$

	A	B	C	D	E	F
1						
2						
3						
4						
5						6
6						

40   *Linear Equations*

Name _____   Date _____   Puzzle #32

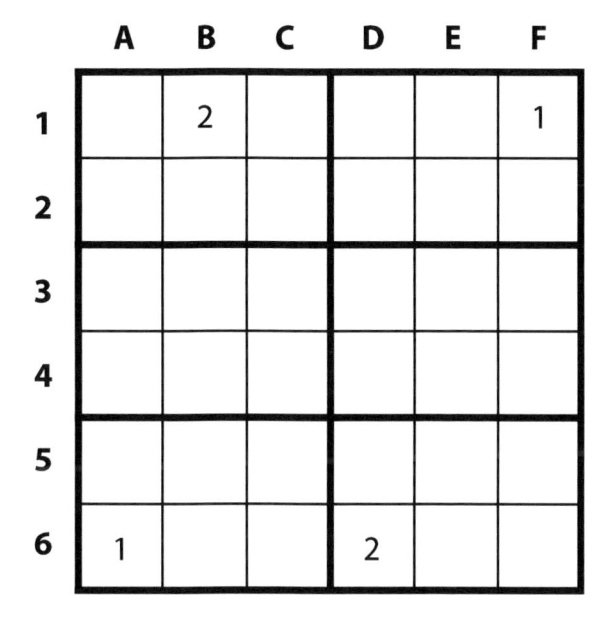

ANOTHER APPLICATION FOR POINT–SLOPE FORM

Point–slope form is also useful in writing an equation when only two points passing through the line are known. First, find the slope of the line using the slope formula, ($m = \frac{y_2-y_1}{x_2-x_1}$). Then use one of the points and the slope in the point–slope equation and solve for y.

Here is an example using the points (3,7) and (-2,-3).
First, find the slope:

$$m = \frac{-3-7}{-2-3} = \frac{-10}{-5} = 2$$

Now, use one of the points and the slope in the point-slope equation:
  y – 7 = 2 (x – 3)
  y – 7 = 2x – 6
y – 7 + 7 = 2x – 6 + 7
      y = 2x + 1

Directions: Given two points passing through a line, write an equation in slope-intercept form. Find your answer in the box on the right. Place the numbers in the appropriate cells of the Sudoku grid, and then solve the puzzle. Show all work!

(A3) (1,-1) (0,1)

(B4) (-4,-4) (2,-1)

(C1) (-3,8) (4,1)

(3) $y = \frac{3}{2}x + 1$
(6) $y = 3x$
(2) $y = -2x + 1$
(1) $y = \frac{1}{2}x - 2$
(4) $y = -x + 5$
(5) $y = -5x - 4$

(E3) (2,4) (-2,-2)

(E6) (-3,11) (0,-4)

(F4) (-2,-6) (1,3)

	A	B	C	D	E	F
1		2				1
2						
3						
4						
5						
6	1			2		

Linear Equations

Name _____ Date _____ Puzzle #33

GRAPHING INEQUALITIES

Graphing the solution of inequalities is similar to graphing the solution of linear equations. First, use the slope and y-intercept to graph the line. If the inequality involves $\leq$ or $\geq$, use a full line, because the points on the line are considered part of the solution. If the inequalities involves < or >, use a dotted or broken line, because the points on the line are not considered in the solution.

The final step is to shade the correct side of the line. To determine which side to shade, use a "test point". Choose any point on either side of the line as a test point (point (0,0) is easiest, as long as the line doesn't run through the origin).

Here is an example using the inequality **2x – y > 3**:

In slope-intercept form,
2x – y > 3 becomes y < 2x – 3

Next, graph the line with m = 2 and b = -3. Because the inequality involves < or >, a broken line is used. Use (0,0) as a test point in the original inequality. This results in 0 < - 3, which is not true. Therefore, the side opposite of (0,0) is shaded.

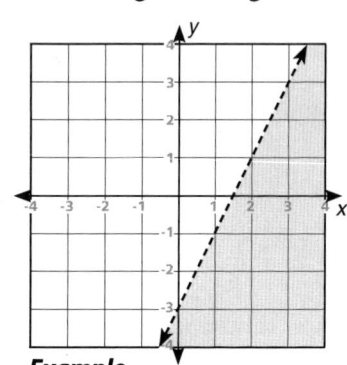

Example

Directions: Match the four equations with graphs below. Place the corresponding values in the appropriate cells of the Sudoku grid, and then complete the puzzle. Show all work!

(A6) $-4y - 8 > 4x$ **(B1)** $2x + 3y \leq 3$ **(F1)** $\frac{2}{5} y + \frac{4}{5} = \frac{1}{5} x$ **(E3)** $-2y + 2 + 4x \leq 0$

#2

#4

#5

#1

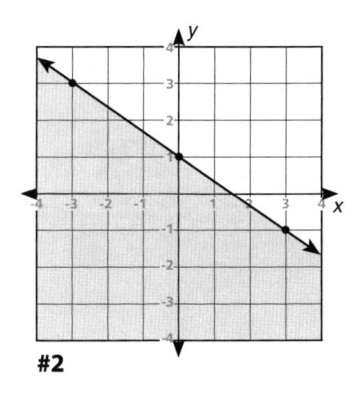

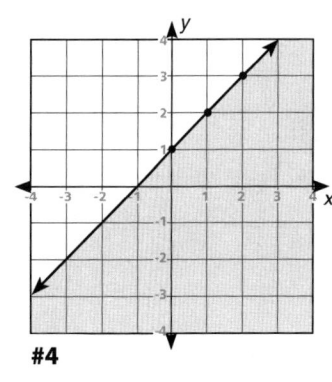

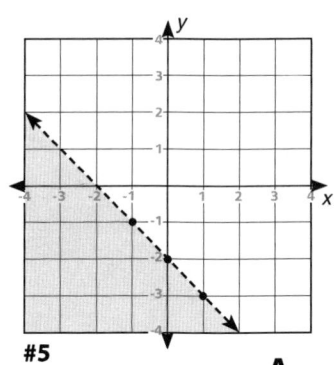

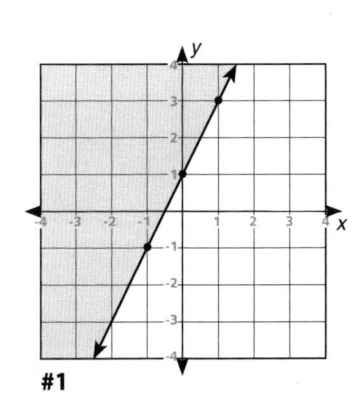

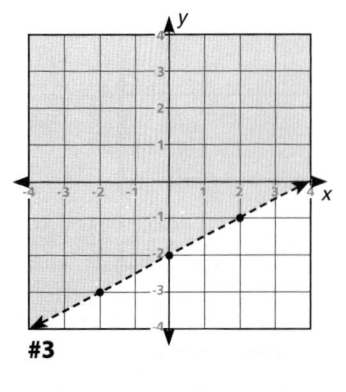

#3

	A	B	C	D	E	F
1				6		
2						
3			2			
4		3		4		
5						
6			3		2	

42   *Linear Equations*

ANSWER KEY

Puzzle #1

	A	B	C	D	E	F	G	H	I
1	-7	-8	1	2	9	-5	6	-3	-4
2	2	9	3	1	4	6	5	-7	8
3	4	6	5	3	7	8	9	2	1
4	5	1	9	8	2	7	3	4	6
5	6	2	4	5	1	3	7	8	-9
6	8	3	7	-4	6	9	1	5	2
7	3	5	-2	6	8	1	4	9	7
8	9	4	-6	7	5	2	8	1	-3
9	-1	-7	8	-9	3	4	2	-6	-5

Puzzle #2

	A	B	C	D	E	F	G	H	I
1	4	8	3	1	-9	6	5	7	2
2	5	2	6	4	3	7	8	-9	1
3	1	9	-7	8	-2	5	4	3	6
4	7	-5	-4	6	-1	2	3	-8	9
5	-3	6	9	-5	8	-4	1	2	7
6	2	-1	8	9	-7	3	-6	4	-5
7	-9	4	1	2	5	8	-7	6	3
8	8	-7	5	3	6	-9	2	1	-4
9	6	3	-2	7	-4	1	9	-5	8

Puzzle #3

	A	B	C	D	E	F	G	H	I
1	8	7	2	6	4	9	-3	-1	-5
2	1	3	4	2	5	-8	9	6	-7
3	6	5	9	1	-3	7	4	8	2
4	5	9	-6	-8	2	4	7	3	1
5	3	-1	7	-9	6	5	2	-4	8
6	4	2	-8	7	1	-3	-5	9	6
7	2	6	3	5	-9	1	8	7	4
8	-7	4	5	-3	8	6	1	-2	9
9	-9	-8	1	4	-7	2	6	5	3

Puzzle #4

	A	B	C	D	E	F	G	H	I
1	-1	-7	-6	4	2	-9	8	5	3
2	5	9	2	8	-1	3	7	4	6
3	3	8	-4	7	6	-5	2	9	1
4	2	1	9	6	5	-8	-3	7	4
5	7	-4	3	1	9	2	6	-8	5
6	8	6	5	3	7	4	9	1	2
7	4	3	1	-9	8	6	-5	2	7
8	6	5	-8	2	4	7	1	-3	9
9	9	2	7	-5	3	1	4	-6	8

43

Puzzle #5

	A	B	C	D	E	F	G	H	I
1	2	4	8	9	7	1	-5	6	3
2	3	6	-1	8	5	-4	2	7	-9
3	9	7	5	6	3	2	1	4	8
4	7	1	4	2	9	5	3	8	6
5	5	3	9	7	8	6	4	1	2
6	8	2	6	4	-1	-3	9	5	7
7	6	8	3	1	4	9	7	2	5
8	4	9	2	5	6	7	8	-3	1
9	1	5	7	3	2	8	6	9	4

Puzzle #6

	A	B	C	D	E	F
1	6	3	1	2	4	5
2	5	2	4	3	1	6
3	1	6	2	5	3	4
4	4	5	3	6	2	1
5	3	4	6	1	5	2
6	2	1	5	4	6	3

Puzzle #7

	A	B	C	D	E	F
1	3	2	6	5	1	4
2	1	4	5	3	6	2
3	2	1	3	4	5	6
4	5	6	4	2	3	1
5	6	5	2	1	4	3
6	4	3	1	6	2	5

Puzzle #8

	A	B	C	D	E	F
1	6	4	2	3	5	1
2	5	1	3	2	6	4
3	3	6	1	4	2	5
4	2	5	4	1	3	6
5	1	2	6	5	4	3
6	4	3	5	6	1	2

Puzzle #9

	A	B	C	D	E	F
1	2	3	6	1	4	5
2	4	5	1	6	2	3
3	6	4	3	5	1	2
4	1	2	5	3	6	4
5	5	6	2	4	3	1
6	3	1	4	2	5	6

Puzzle #10

	A	B	C	D	E	F
1	1	6	4	3	5	2
2	3	5	2	1	6	4
3	4	3	5	2	1	6
4	2	1	6	4	3	5
5	6	4	3	5	2	1
6	5	2	1	6	4	3

Puzzle #11

	A	B	C	D	E	F	G	H	I
1	7	-9	-4	5	8	3	2	1	6
2	8	-3	6	7	2	1	4	-9	5
3	-1	5	2	-4	-9	6	7	8	3
4	3	7	1	2	6	4	9	5	8
5	5	2	9	3	7	8	1	6	4
6	6	4	8	9	1	-5	3	2	7
7	9	6	7	8	3	2	5	4	1
8	4	8	3	1	5	-9	6	7	2
9	2	1	5	6	4	7	8	3	9

Puzzle #12

	A	B	C	D	E	F	G	H	I
1	1	4	9	3	5	2	6	7	8
2	2	6	3	1	7	8	9	5	4
3	7	8	5	6	4	9	3	2	1
4	3	2	1	9	8	4	5	6	7
5	6	7	8	5	2	3	1	4	9
6	5	9	4	7	6	1	8	3	2
7	8	5	6	2	9	7	4	1	3
8	4	1	2	8	3	6	7	9	5
9	9	3	7	4	1	5	2	8	6

Puzzle #13

	A	B	C	D	E	F
1	4	5	2	6	1	3
2	3	6	1	5	2	4
3	5	1	-4	2	3	6
4	6	2	3	-1	4	5
5	2	4	6	3	5	1
6	1	3	-5	4	6	2

Puzzle #14

	A	B	C	D	E	F
1	6	1	-3	4	5	2
2	4	2	5	6	3	1
3	5	3	2	1	-6	4
4	1	6	4	5	2	3
5	2	5	1	3	4	-6
6	3	4	6	2	1	5

Puzzle #15

	A	B	C	D	E	F
1	6	5	2	4	3	1
2	3	1	4	2	6	5
3	5	4	3	6	1	2
4	1	2	6	3	5	4
5	2	3	5	1	4	-6
6	4	6	1	5	2	3

Puzzle #16

	A	B	C	D	E	F
1	5	3	1	6	4	2
2	6	4	2	5	3	1
3	4	2	5	-3	1	6
4	3	1	6	4	2	5
5	2	5	3	1	6	4
6	1	6	4	2	5	-3

Puzzle #17

	A	B	C	D
1	4	2	3	1
2	3	1	4	-2
3	2	3	1	4
4	1	4	2	3

46

Puzzle #18

	A	B	C	D	E	F
1	6	2	5	3	-4	1
2	1	4	3	2	6	5
3	5	1	2	6	3	4
4	3	6	4	1	5	2
5	2	5	6	4	1	3
6	4	3	1	5	2	6

Puzzle #19

	A	B	C	D	E	F
1	1	3	2	5	4	6
2	4	5	6	1	2	3
3	3	2	5	6	1	4
4	6	4	1	3	5	2
5	2	1	3	4	6	5
6	5	6	4	2	3	1

Puzzle #20

	A	B	C	D	E	F
1	2	3	1	5	6	4
2	5	4	6	1	3	2
3	3	1	4	6	2	5
4	6	5	2	3	4	1
5	1	2	3	4	5	6
6	4	6	5	2	1	3

Puzzle #21

	A	B	C	D	E	F
1	-3	-1	6	4	-2	5
2	2	4	5	6	1	3
3	6	3	2	5	4	1
4	4	5	-1	3	6	-2
5	1	6	3	2	5	4
6	5	2	4	1	3	6

Puzzle #22

	A	B	C	D	E	F	G	H	I
1	8	4	1	6	7	3	2	5	9
2	9	2	3	8	4	5	6	7	1
3	7	5	6	9	2	1	3	-4	8
4	2	6	4	-7	5	9	1	8	3
5	5	8	9	3	1	6	4	2	7
6	1	3	7	-2	8	-4	9	6	5
7	6	7	8	1	3	2	5	9	4
8	3	9	5	4	6	7	8	-1	2
9	4	1	2	5	9	8	7	3	6

Puzzle #23

	A	B	C	D	E	F
1	6	4	1	2	5	3
2	2	3 or -3	5	4	6	1 or -1
3	3	5	6	1	4 or -4	2
4	4	1	2	5	3	6
5	1	6	4	3 or -3	2	5
6	5	2	3	6	1	4

Puzzle #24

	A	B	C	D	E	F
1	6	3	-5	4	1	2
2	1	4	2	5	3	6
3	4	6	3	2	5	1
4	5	2	1	6	4	3
5	3	5	6	1	2	4
6	-2	1	4	3	6	5

Puzzle #25

	A	B	C	D	E	F
1	-1	2	4	6	5	3
2	5	6	3	1	4	2
3	6	4	1	3	2	5
4	2	3	5	4	6	1
5	4	1	2	5	3	6
6	3	-5	6	2	1	4

48

Puzzle #26

	A	B	C	D	E	F
1	3	2	4	5	6	1
2	5	1	6	2	3	4
3	2	6	5	4	1	3
4	1	4	3	6	2	5
5	4	3	2	1	5	6
6	6	5	1	3	4	2

Puzzle #27

	A	B	C	D	E	F
1	6	1	3	5	2	4
2	4	2	5	6	3	1
3	1	5	6	2	4	3
4	2	3	4	1	6	5
5	3	6	1	4	5	2
6	5	4	2	3	1	6

Puzzle #28

	A	B	C	D	E	F
1	4	5	6	1	2	3
2	1	3	2	5	4	6
3	3	4	5	2	6	1
4	6	2	1	4	3	5
5	5	6	4	3	1	2
6	2	1	-3	6	5	-4

Puzzle #29

	A	B	C	D	E	F
1	3	5	1	6	4	2
2	6	2	4	5	3	1
3	1	3	2	4	5	6
4	4	6	5	1	2	3
5	5	1	3	2	6	4
6	2	4	6	3	1	5

Puzzle #30

	A	B	C	D	E	F
1	3	4	5	6	2	1
2	2	6	1	3	5	4
3	4	1	3	5	6	-2
4	6	5	-2	1	4	3
5	5	3	4	2	1	6
6	1	2	6	4	-3	5

Puzzle #31

	A	B	C	D	E	F
1	2	3	5	6	1	4
2	6	4	1	5	3	2
3	4	1	2	3	6	5
4	3	5	6	4	2	1
5	5	2	3	1	4	6
6	1	6	4	2	5	3

Puzzle #32

	A	B	C	D	E	F
1	5	2	4	3	6	1
2	6	3	1	5	4	2
3	2	4	6	1	3	5
4	3	1	5	4	2	6
5	4	5	2	6	1	3
6	1	6	3	2	5	4

Puzzle #33

	A	B	C	D	E	F
1	4	2	1	6	5	3
2	3	5	6	2	4	1
3	6	4	2	3	1	5
4	1	3	5	4	6	2
5	2	1	4	5	3	6
6	5	6	3	1	2	4

NOTES

REFERENCES

Adelman, C. *Answers in the tool box: Academic intensity, attendance patterns, and bachelor's degree attainment.* Washington, DC: U.S. Department of Education, Office of Educational Research and Improvement, 1999.

Business Higher Education Forum. *A Commitment to America's Future: Responding to the Crisis in Mathematics and Science Education - The Main Report.* Washington, DC, 2005.
http://www.bhef.com/www/publications/documents/commitment_future_05.pdf

Evan, A., Gray, T., & Olchefske, J. *The gateway to student success in mathematics and science.* Washington, DC: American Institutes for Research, 2006.

National Mathematics Advisory Panel, *Final Report.* Washington, DC, 2008.

Williams, Tony. *100 Algebra Workouts and Practical Teaching Tips.* Dayton: Teaching and Learning Company, 2008.

Williams, Tony. *100 Math Workouts and Practical Teaching Tips.* Dayton: Teaching and Learning Company, 2008.